Name _____ Class _____

Skills Worksheet

Directed Reading

Section: The Theory of Evolution by Natural Selection

In the space provided, write the letter of the term or phrase that best completes each statement or best answers each question.

_____ 1. Some individuals of a population or species are better suited to
   a. evolve than other individuals.
   b. survive and reproduce.
   c. become extinct.
   d. achieve punctuated equilibrium.

_____ 2. Charles Darwin is credited with the theory of
   a. evolution by natural selection.
   b. populations.
   c. evolution by gradualism.
   d. gravitation.

_____ 3. In science, evolution is referred to as
   a. mere speculation.
   b. an explanation of why species become extinct.
   c. change over time.
   d. an explanation for the rates of population growth.

_____ 4. Darwin learned that there were resemblances between the plants and animals of South America and
   a. Ecuador.
   b. the Galápagos Islands.
   c. Australia.
   d. the English countryside.

Read the question, and write your answer in the space provided.

5. What was the mechanism for evolution proposed by Jean Baptiste Lamarck?

   _____

   _____

In the space provided, explain how the terms in each pair differ in meaning.

6. population, species

   _____

   _____

Copyright © by Holt, Rinehart and Winston. All rights reserved.

Holt Biology   1   The Theory of Evolution

Name _____ Class _____ Date _____

Directed Reading continued

7. adaptation, natural selection

_____

_____

Complete each statement by underlining the correct term or phrase in the brackets.

8. Traits of individuals best suited to survive will become [more / less] common in each new generation.

9. [Genes / Natural selection] is (are) responsible for inherited traits.

10. [Natural selection / Genes] cause(s) the frequency of certain alleles in a population to vary over time.

11. [Isolation / Extinction] is the condition in which two populations of the same species cannot breed with one another.

12. Generally, when the individuals of two related populations can no longer breed with one another, the two populations are considered to be different [organisms / species].

Skills Worksheet

Directed Reading

Section: Evidence of Evolution

In the space provided, write the letter of the term or phrase that best completes each statement or best answers each question.

_____ 1. One hypothesized link between modern whales and hoofed mammals are
   a. fish.
   b. mesonychids.
   c. penguins.
   d. alligators.

_____ 2. Links between major classes of vertebrates have been established by
   a. radiometric dating.
   b. inherited traits.
   c. the fossil record.
   d. patterns of development.

_____ 3. Most scientists agree that
   a. Earth is 4.5 billion years old.
   b. Earth has supported life for most of its history.
   c. Living organisms share ancestry with earlier, simpler life-forms.
   d. All of the above

_____ 4. The fossil record
   a. proves the existence of every species that has ever lived.
   b. cannot show patterns of development from ancestors to descendants.
   c. shows strong evidence that evolution takes place.
   d. cannot show change over time in species.

_____ 5. A paleontologist is a scientist who studies
   a. fossils.
   b. theories.
   c. anatomy and development.
   d. biological molecules.

_____ 6. Fossils form when organisms are rapidly buried in
   a. sand.
   b. grass.
   c. leaf litter.
   d. fine sediment.

Directed Reading continued

In the space provided, write the letter of the description that best matches the term or phrase.

_____ 7. vestigial structures

_____ 8. homologous structures

_____ 9. vertebrate embryos

a. structures that have no use or little use and are evidence of an organism's evolutionary past

b. pharyngeal pouches and tails are evidence of evolution

c. structures that share a common ancestry

Complete each statement by writing the correct term or phrase in the space provided.

10. Species that diverged recently have _____ genetic differences than those species that are not closely related.

11. There is (are) _____ difference(s) between the amino acid sequences of the hemoglobin in humans and the hemoglobin in gorillas.

12. There is (are) _____ difference(s) between the amino acid sequences of the hemoglobin in humans and the hemoglobin in frogs.

13. There is (are) _____ difference(s) between the amino acid sequences of the hemoglobin in humans and the hemoglobin in rhesus monkeys.

14. Scientists are able to determine the exact amino acid sequence of a(n) _____ .

Read each question, and write your answer in the space provided.

15. How do scientists estimate the number of nucleotide changes that have taken place in a gene since two species diverged from a common ancestor?

16. How does comparison of amino acid differences between species provide evidence of evolution?

Name _____ Class _____ Date _____

Skills Worksheet

Directed Reading

Section: Examples of Evolution

Complete each statement by writing the correct term or phrase in the space provided.

1. The _____ presents many different challenges to an individual's ability to reproduce.

2. Organisms tend to produce more _____ than their environment can support.

3. All populations have genetic _____ .

4. Individuals of a species often _____ with one another to survive.

5. Individuals within a population that are better able to cope with the challenges of their environment tend to leave _____ offspring than those less suited to the environment.

Read each question, and write your answer in the space provided.

6. What is antibiotic resistance?

   _____

   _____

7. Describe the study conducted by Peter and Rosemary Grant.

   _____

   _____

8. What was the environmental challenge in the Grants' study?

   _____

   _____

9. What was the effect of natural selection on beak size in the Grants' study?

   _____

   _____

Copyright © by Holt, Rinehart and Winston. All rights reserved.

Holt Biology — The Theory of Evolution

Name _____ Class _____ Date _____

Directed Reading *continued*

In the space provided, explain how the terms in each pair differ in meaning.

10. divergence, speciation

_____

_____

11. subspecies, species

_____

_____

Complete each statement by writing the correct term or phrase in the space provided.

12. In frogs, different mating seasons are a(n) _____ to reproduction.

13. Reproductive _____ is the inability of formerly interbreeding groups to mate or produce fertile offspring.

14. The way that natural selection leads to the formation of new _____ has been thoroughly documented.

Name _____ Class _____ Date _____

Skills Worksheet

Active Reading

Section: The Theory of Evolution by Natural Selection

Read the passage below. Then answer the questions that follow.

Darwin realized that Malthus's hypotheses about human populations apply to all species. Every organism has the potential to produce many offspring during its lifetime. In most cases, however, only a limited number of those offspring survive to reproduce. Adding Malthus's view to what he saw on his voyage and to his own experiences in breeding domestic animals, Darwin made a key association: *Individuals that have physical or behavioral traits that better suit their environment are more likely to survive and will reproduce more successfully than those that do not have such traits.* Darwin suggested that by surviving long enough to reproduce, individuals have the opportunity to pass on their favorable characteristics to offspring. In time, these favorable characteristics will increase in a population, and the nature of the population will gradually change. Darwin called this process by which populations change in response to their environment **natural selection.**

SKILL: READING EFFECTIVELY

Read each question, and write your answer in the space provided.

1. Based on the first three sentences of this passage, what can the reader infer was Malthus's idea about the human population?

_____

_____

_____

2. What real-life experiences of his own did Darwin reflect upon when considering Malthus's ideas about human populations?

_____

_____

_____

3. According to Darwin, what causes the nature of a population to change?

_____

_____

_____

Copyright © by Holt, Rinehart and Winston. All rights reserved.

Holt Biology   The Theory of Evolution

Name _____ Class _____ Date _____

Active Reading continued

Read this second passage below. Then answer the questions that follow.

Scientists now know that genes are responsible for inherited traits. Therefore, certain forms of a trait become more common in a population because more individuals in the population carry the alleles for those forms. In other words, natural selection causes the frequency of certain alleles in a population to increase or decrease over time. Mutations and the recombination of alleles that occurs during sexual reproduction provide endless sources of new variations for natural selection to act upon.

SKILL: READING EFFECTIVELY

Read each question, and write your answer in the space provided.

4. What controls inherited traits?

   _____

   _____

5. What causes a particular trait to become more common in a population?

   _____

   _____

6. What two events cause new variations of traits in a population?

   _____

   _____

Skills Worksheet

Active Reading

Section: Evidence of Evolution

Read the passage below. Then answer the questions that follow.

The fossil record, and thus the record of the evolution of life, is not complete. Many species have lived in environments where fossils do not form. Most fossils form when organisms and traces of organisms are rapidly buried in fine sediments deposited by water, wind, or volcanic eruptions. The environments that are most likely to cause fossil formation are wet lowlands, slow-moving streams, lakes, shallow seas, and areas near volcanoes that spew out volcanic ash. The chances that organisms living in upland forests, mountains, grasslands, or deserts will die in just the right place to be buried in sediments and fossilized are very low. Even if an organism lives in an environment where fossils can form, the chances are slim that its dead body will be buried in sediment before it decays. For example, the organism may be eaten and scattered by scavengers.

READING EFFECTIVELY

Read each question, and write your answer in the space provided.

1. Why is the fossil record incomplete?

_____
_____
_____

2. Where do fossils form?

_____
_____
_____
_____
_____

3. In areas where fossils form, why don't all organisms that die become fossilized?

_____
_____

Copyright © by Holt, Rinehart and Winston. All rights reserved.

Holt Biology　　　　　　　　　　　　　　　The Theory of Evolution

Name _____ Class _____ Date _____

Skills Worksheet

Active Reading

Section: Examples of Evolution

The figure below shows beak-size variations in finches. Using the information contained in the figure, answer each question in the space provided.

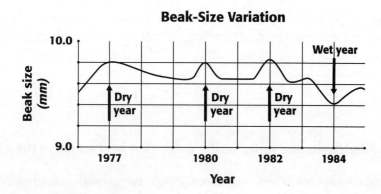

SKILL: INTERPRETING GRAPHICS

1. The title of a graph indicates the type of information it contains. What is the title of the graph shown? Based on this title, what type of information can an observer expect to find?

   _____
   _____
   _____
   _____

2. Read the label on the horizontal axis of the graph. What information is plotted along this axis?

   _____
   _____

3. What intervals are used on the horizontal axis?

   _____
   _____

4. Read the label on the vertical axis of the graph. What information is plotted along this axis?

   _____

Copyright © by Holt, Rinehart and Winston. All rights reserved.

Holt Biology   The Theory of Evolution

Name _____ Class _____ Date _____

Active Reading *continued*

5. What units are used on the vertical axis?
_____
_____

6. Based on the data shown, what effect does a dry year have on beak size?
_____
_____

7. Based on the data shown, what effect does a wet year have on beak size?
_____
_____

In the space provided, write the letter of the dates that best answers the question.

_____ **8.** During which two years was the average finch beak size nearly the same?
   a. 1976 and 1982
   b. 1977 and 1979
   c. 1979 and 1981
   d. 1980 and 1983

Name _____ Class _____ Date _____

Skills Worksheet

Vocabulary Review

In the space provided, write the letter of the term or phrase that best completes each statement or best answers each question.

_____ 1. The process in which organisms with traits well suited to an environment are more likely to survive and to produce offspring is
   a. trait mechanisms.
   b. origin of species.
   c. genetic principles.
   d. natural selection.

_____ 2. In biology, all of the individuals of a species that live together in one place at one time are called a
   a. population.
   b. community.
   c. half-life.
   d. habitat.

_____ 3. A change in the genetic makeup of species over time is called
   a. radioactive dating.
   b. evolution.
   c. camouflage.
   d. natural selection.

_____ 4. The process by which a species becomes better suited to its environment is
   a. industrialization.
   b. not an advantage.
   c. adaptation.
   d. destructive to its survival.

_____ 5. Structures that share a common ancestry or are similar because they are modified versions of structures from a common ancestor are
   a. not related.
   b. homologous.
   c. not homologous.
   d. young in origin.

_____ 6. Structures with no function that are remnants of an organism's evolutionary past are
   a. not visible on organisms.
   b. young in origin.
   c. vestigial.
   d. useful to the organism.

_____ 7. The accumulation of differences between species or populations is called
   a. gradualism.
   b. punctuated equilibrium.
   c. divergence.
   d. observational species.

_____ 8. The hypothesis that evolution of a species occurs in periods of rapid change separated by periods of little or no change is called
   a. divergence.
   b. gradualism.
   c. isolation.
   d. punctuated equilibrium.

_____ 9. Populations of the same species that differ genetically because they have adapted to different living conditions are
   a. observational species.
   b. different species.
   c. subspecies.
   d. conditional races.

Copyright © by Holt, Rinehart and Winston. All rights reserved.

Holt Biology — The Theory of Evolution

Name _____ Class _____ Date _____

Vocabulary Review *continued*

_____ 10. The hypothesis that the evolution of different species occurs at a slow constant rate is called
   a. punctuated equilibrium.
   b. gradualism.
   c. divergence.
   d. transitionism.

_____ 11. The condition in which two populations of the same species CANNOT breed with one another is called reproductive
   a. infertility.
   b. extinction.
   c. isolation.
   d. selection.

_____ 12. When a species permanently disappears, the species is said to be
   a. extinct.
   b. isolated.
   c. mutated.
   d. eliminated.

_____ 13. Antibiotic resistance in bacteria is called
   a. natural selection.
   b. gradualism.
   c. divergence.
   d. speciation.

_____ 14. The process by which new species form is called
   a. biological change.
   b. reproduction.
   c. speciation.
   d. divergence.

_____ 15. The inability of formerly interbreeding groups to mate or produce fertile offspring is called
   a. sterility.
   b. divergence.
   c. reproductive isolation.
   d. extinction.

_____ 16. A scientist who studies fossils is called a(n)
   a. archaeologist.
   b. ecologist.
   c. paleontologist.
   d. biologist.

_____ 17. In Grants' study, the effect of weather on the size of the finch's beak is an example of
   a. isolation.
   b. natural selection.
   c. gradualism.
   d. fossilization.

_____ 18. Biological molecules that are considered evidence for evolution include
   a. DNA.
   b. amino acids.
   c. proteins.
   d. All of the above

Skills Worksheet

Science Skills

APPLYING INFORMATION

Darwin stated that evolution occurs through natural selection. The key factor is the environment. The environment "selects" which organisms will survive and reproduce. Traits possessed by organisms successful at survival and reproduction are more likely to be transmitted to the next generation. These traits will, therefore, become common.

Read the following information about the elephant population of Queen Elizabeth National Park in Uganda, Africa. Then use the table on the next page to answer questions 1–5.

Normally, nearly all African elephants, male and female, have tusks. In 1930, only 1 percent of the elephant population in Queen Elizabeth National Park was tuskless because of a rare genetic mutation. Food was fairly plentiful, and by 1963, there were 3,500 elephants in the park. In the 1970s, a civil war began in Uganda. Much of the wildlife was killed for food, and poachers killed elephants for their ivory tusks. By 1992, the elephant population had dropped to about 200. But by 1998, the population had increased to 1,200. A survey in 1998 revealed that as many as 30 percent of the adult elephants did not have tusks. Ugandan wildlife officials also noted a decline in poaching.

Name _____ Class _____ Date _____

Science Skills continued

In the space provided in the table below, explain how each of the given principles of natural selection applies to the situation described on the previous page.

The Process of Natural Selection

Principles	Applications
All species have genetic variation.	1. _____
Living things face many challenges in the struggle to exist.	2. _____
Individuals of species often compete with one another to survive.	3. _____
Individuals that are better able to cope with the challenges of their environment tend to leave more offspring than those less suited.	4. _____
The characteristics of the individuals best suited to a particular environment tend to increase in a population over time.	5. _____

Name _____ Class _____ Date _____

Skills Worksheet

Concept Mapping

Using the terms and phrases provided below, complete the concept map showing the theory of evolution.

embryonic development   homologous structures   proteins
finch beaks   natural selection   punctuated equilibrium
fossils   nucleic acids   vestigial structures
gradualism   antibiotic environment

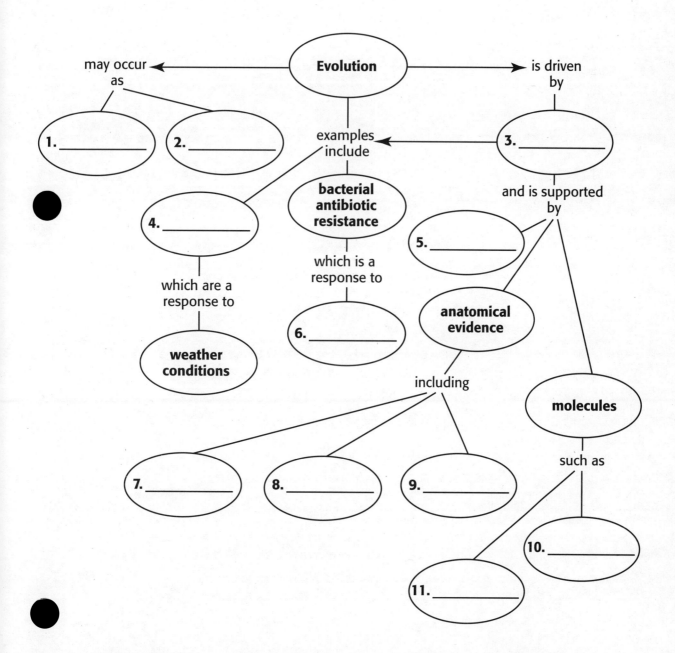

Copyright © by Holt, Rinehart and Winston. All rights reserved.

Holt Biology — The Theory of Evolution

Name _____ Class _____ Date _____

Skills Worksheet

Critical Thinking

Work-Alikes

In the space provided, write the letter of the term or phrase that best describes how each numbered item functions.

_____ 1. natural selection

_____ 2. punctuated equilibrium

_____ 3. fossil

_____ 4. HMS *Beagle* voyage

a. running short sprints with long rest periods in between

b. short trees might not be as likely to be hit by lightning in a rainstorm

c. naturalist's field expedition

d. a clue in a detective story

Cause and Effect

In the space provided, write the letter of the term or phrase that best matches each cause or effect given below.

Cause	Effect
5. individuals best suited to survive have the most offspring	_____
6. _____	common ancestry indicated
7. _____	antibiotic-resistant bacteria predominate
8. two populations of same species become different	_____
9. environment directs change	_____
10. _____	subspecies
11. _____	reproductive isolation
12. nucleotide substitution	_____

a. changes in species are kept through natural selection

b. their traits will be passed on to future generations

c. unable to breed with one another

d. changes in amino acid sequence in a protein

e. same species become different in different environments

f. members of same species reproduce at different times

g. homologous structures

h. mutant bacteria favored to survive antibiotic treatment

Copyright © by Holt, Rinehart and Winston. All rights reserved.

Critical Thinking continued

Linkages

In the spaces provided, write the letters of the two terms or phrases that are linked together by the term or phrase in the middle. The choices can be placed in any order. Some choices may be used more than once.

13. _____ South American species must have migrated to the Galápagos Islands _____

14. _____ orderly patterns of fossils are seen _____

15. _____ there are few differences in sequences _____

16. _____ structures develop at different rates _____

a. ages can be determined
b. amino acid sequences analyzed
c. plants and animals of South America and Galápagos Islands resemble one another
d. vertebrate embryos develop a tail, buds, and pouches
e. in humans, the tail disappears
f. South American species changed
g. shared common ancestry more recently
h. radiometric dating arranges fossils

Analogies

An analogy is a relationship between two pairs of terms or phrases written as a : b :: c : d. The symbol : is read as "is to," and the symbol :: is read as "as." In the space provided, write the letter of the pair of terms or phrases that best completes the analogy shown.

_____ 17. amino acid differences : divergence of species ::
   a. atoms : years
   b. potassium : uranium
   c. rocks : half-life
   d. radioactive decay : age of rock

_____ 18. gradualism : punctuated equilibrium ::
   a. running : walking
   b. reading : writing
   c. crawling : flying
   d. understanding : confusing

_____ 19. divergence : speciation ::
   a. common ancestors : mating
   b. reproduction : isolation
   c. barriers : reproductive isolation
   d. prevention of mating : reproduction

Name _____ Class _____ Date _____

Skills Worksheet

Test Prep Pretest

In the space provided, write the letter of the term or phrase that best completes each statement or best answers each question.

_____ 1. On the Galápagos Islands, Darwin saw that the plants and animals closely resembled those of the
   a. islands off the coast of North America.
   b. coast of South America.
   c. islands off the coast of Africa.
   d. coast of South Africa.

_____ 2. Which of the following is a factor in natural selection?
   a. Individuals of a species compete with one another to survive.
   b. All species are genetically diverse.
   c. Individuals better able to adapt to changes leave more offspring.
   d. All of the above

_____ 3. When the individuals of two populations can no longer interbreed, the two populations are considered to be
   a. different families.   c. the same species.
   b. different species.   d. unrelated.

_____ 4. The fossil record provides evidence that
   a. older species gave rise to more-recent species.
   b. all species were formed during Earth's formation and have changed little since then.
   c. the fossilized species have no connection to today's species.
   d. fossils cannot be dated.

_____ 5. Comparing human hemoglobin with the hemoglobin of gorillas, mice, chickens, and frogs reveals that humans have the fewest amino acid differences with
   a. gorillas.   c. chickens.
   b. mice.   d. frogs.

_____ 6. Individuals that are better able to cope with the challenges of their environment tend to
   a. decrease in population over time.
   b. leave more offspring than those more suited to the environment.
   c. leave fewer offspring than those less suited to the environment.
   d. leave more offspring than those less suited to the environment.

_____ 7. Which factor does NOT play a role in determining the beak size of Galápagos finches?
   a. amount of food available   c. size of the bird
   b. seed size   d. weather

Name _____ Class _____ Date _____

Test Prep Pretest *continued*

_____ 8. Members of different subspecies
   a. are considered to be different species.
   b. differ genetically because of adaptations for different living conditions.
   c. can no longer interbreed successfully.
   d. will never diverge to become different species.

Questions 9 and 10 refer to the figures below.

Chicken embryo **Human embryo**

(Chicken embryo labeled: Pharyngeal pouch, Bony tail)
(Human embryo labeled: Pharyngeal pouch)

_____ 9. Which of the following statements best reflects the evolutionary importance of the figures above?
   a. New genetic instructions have been disregarded in the evolution of vertebrates.
   b. Early in development, vertebrate embryos show no evidence of common ancestry.
   c. The evolutionary history of organisms is seen in the way embryos develop.
   d. All adult vertebrates retain pharyngeal pouches.

_____ 10. Which of the following statements is NOT true about the vertebrate embryos shown above?
   a. Each embryo develops a tail.
   b. Each embryo has buds that become limbs.
   c. Each embryo has pharyngeal pouches.
   d. Each embryo has fur.

Test Prep Pretest continued

Complete each statement by writing the correct term or phrase in the space provided.

11. Over time, change within species leads to the replacement of old species by new species as less successful species become _____ .

12. While on the *Beagle*, Darwin read *Principles of Geology*, which contained a detailed account of _____ theory of evolution.

13. The changing of a species that results in its being better suited to its environment is called _____ .

14. A(n) _____ is a group of individuals that belong to the same species, live in a defined area, and breed with others in the group.

15. The condition in which two populations of the same species are separated from one another is called _____ .

16. Species that shared a common ancestor in the recent past have many _____ _____ or _____ sequence similarities.

17. Given that the forelimbs of all vertebrates share the same basic arrangement of bones, forelimbs are said to be _____ structures.

18. The _____ of individuals who adapt to changing conditions tend to increase over time.

19. The model of evolution in which gradual change leads to species formation over time is called _____ .

20. A whale's pelvic bones are _____ structures because they no longer function like the pelvis of a land vertebrate.

21. Darwin felt that fossils of extinct armadillos that resembled living armadillos were evidence that _____ is a(n) _____ process.

22. The accumulation of differences between groups such as populations, species, and genera is _____ .

Name _____ Class _____ Date _____

Test Prep Pretest *continued*

Read each question, and write your answer in the space provided.

23. What was Lamarck's hypothesis regarding evolution?

_____

_____

_____

24. Briefly explain the importance of Thomas Malthus's essay on the growth of the human population to Darwin's theory of evolution.

_____

_____

_____

25. Briefly summarize the modern version of Darwin's theory of evolution by natural selection.

_____

_____

_____

_____

_____

Assessment Quiz

Section: The Theory of Evolution by Natural Selection

In the space provided, write the letter of the term or phrase that best completes each statement or best answers each question.

_____ 1. The voyage of the *Beagle* took Darwin to
   a. Australia.
   b. the Galápagos Islands.
   c. South America.
   d. All of the above

_____ 2. Darwin believed that Malthus's hypotheses about populations applied
   a. to all species.
   c. to all species except humans.
   b. to only the human population.
   d. to only a specific population.

_____ 3. Variation exists within the genes of every population or species as the result of
   a. environmental factors.
   b. inheritance.
   c. random mutation.
   d. recessive characteristics.

_____ 4. Darwin and Lamarck agreed that changes in species are linked to
   a. genetics.
   b. environmental conditions.
   c. use or disuse of physical features.
   d. acquired characteristics.

_____ 5. Natural selection causes the frequency of favorable alleles to
   a. decrease in a population.
   b. remain constant in a population.
   c. increase in a population.
   d. vary widely in a population.

In the space provided, write the letter of the description that best matches the name(s) of the scientist(s).

_____ 6. Charles Darwin

_____ 7. Charles Lyell

_____ 8. Alfred Russell Wallace

_____ 9. Gould and Eldredge

_____ 10. Jean Baptiste Lamarck

a. proposed a theory of evolution by means of acquired characteristics

b. wrote an essay describing evolution by natural selection

c. *On the Origin of Species by Means of Natural Selection*

d. punctuated equilibrium

e. *Principles of Geology*

Name _____ Class _____ Date _____

Assessment

Quiz

Section: Evidence of Evolution

In the space provided, write the letter of the term or phrase that best completes each statement or best answers each question.

_____ 1. Most scientists agree that
   a. Earth is about 3.5 million years old.
   b. life is new on Earth.
   c. living organisms share ancestry.
   d. intermediate fossils do not exist.

_____ 2. Which organism is most likely to become fossilized?
   a. a clam in sediment
   b. an earthworm in sand
   c. a mouse killed by a fox
   d. a tree-dwelling insect

_____ 3. The fossil record
   a. will never be complete.
   b. provides evidence of evolution.
   c. is a record of Earth's past life-forms.
   d. All of the above

_____ 4. A vestigial structure is one that is
   a. similar to structure in other species.
   b. reduced in size and useless.
   c. an embryological structure.
   d. a characteristic of vertebrate.

_____ 5. Species that share a distant common ancestor
   a. have many amino acid sequence differences.
   b. have few amino acid sequence differences.
   c. have identical nucleotide sequences.
   d. are not represented by the fossil record.

In the space provided, write the letter of the description that best matches the term or phrase.

_____ 6. pterodactyl

_____ 7. crinoid

_____ 8. hemoglobin protein

_____ 9. paleontologist

_____ 10. modern whales

a. early multicellular life-form found in 800 million years old rocks
b. determines the age of fossils by radiometric dating
c. most likely had a hooved mammal ancestor
d. an extinct reptile found in rocks 140–210 million years old
e. compared among species to determine evolutionary relationships

Copyright © by Holt, Rinehart and Winston. All rights reserved.
Holt Biology — The Theory of Evolution

Name _____ Class _____ Date _____

Assessment

Quiz

Section: Examples of Evolution

In the space provided, write the letter of the term or phrase that best completes each statement or best answers each question.

_____ 1. Individuals that have traits that enable them to survive in a given environment can reproduce and
   a. begin the process of speciation.
   b. pass on those traits to their offspring.
   c. slow the process of evolution.
   d. All of the above

_____ 2. Which of the following statements was NOT suggested by Darwin?
   a. Natural selection is the mechanism that drives evolution.
   b. Antibiotic-resistant strains of tuberculosis have evolved by natural selection.
   c. There are many examples of how natural selection has shaped life on Earth.
   d. The number of finches with different beak shapes are changed by natural selection.

_____ 3. Geographic isolation between two populations of a species may lead to
   a. divergence and speciation.
   b. increased mating between the populations.
   c. inclement weather.
   d. unsuitable nesting sites.

_____ 4. The antibiotic used effectively to treat tuberculosis before bacteria became resistant to it is
   a. penicillin.   c. rifampin.
   b. mycobacteria.   d. *rpoB*.

_____ 5. The accumulation of differences between groups is called
   a. natural selection.   c. subspecies.
   b. divergence.   d. reproductive isolation.

In the space provided, write the letter of the process that best matches the statement on the left. You will use each letter more than one time.

_____ 6. All populations have genetic variation.

_____ 7. A separated population begins to accumulate genetic differences.

_____ 8. The environment presents challenges to successful reproduction.

a. an element in the process of natural selection

b. a step in the process of species formation

Copyright © by Holt, Rinehart and Winston. All rights reserved.
Holt Biology   The Theory of Evolution

Name _____ Class _____ Date _____

Quiz *continued*

_____ **9.** Individuals tend to produce more offspring than the environment can support.

_____ **10.** A population diverges sufficiently to become a subspecies.

Name _____ Class _____ Date _____

Assessment

Chapter Test

The Theory of Evolution

In the space provided, write the letter of the term or phrase that best completes each statement or best answers each question.

_____ 1. The evolution of beak sizes in Galápagos finches is a response to
   a. how finches use their beaks.
   b. the types of seeds available.
   c. whether the populations interbreed.
   d. the nutritional content of the seeds.

_____ 2. According to Darwin, evolution occurs
   a. in response to use or disuse of a characteristic.
   b. by punctuated equilibrium.
   c. by natural selection.
   d. within an individual's lifetime.

_____ 3. The hypothesis that evolution occurs at a rapid rate, separated by periods of no change,
   a. was supported by Darwin.
   b. is known as punctuated equilibrium.
   c. is supported by many transitional forms in the fossil record.
   d. was proposed by Lyell.

_____ 4. The traits of individuals best adapted to survive become more common in each new generation because
   a. offspring without those traits do not survive.
   b. the alleles responsible for those traits increase through natural selection.
   c. those individuals do not breed.
   d. natural selection does not affect well-adapted individuals.

_____ 5. Natural selection causes
   a. changes in the environment.
   b. plants and animals to produce more offspring than can survive.
   c. changes in the frequency of certain alleles in a population.
   d. All of the above

_____ 6. In the study of the bacteria that cause tuberculosis, scientists have learned that
   a. the bacteria have become more sensitive to antibiotics due to evolution.
   b. patients are allergic to antibiotic treatment.
   c. the bacteria have become resistant to antibiotics due to natural selection.
   d. patients frequently die from the antibiotic treatment.

Copyright © by Holt, Rinehart and Winston. All rights reserved.
Holt Biology — The Theory of Evolution

Name _____ Class _____ Date _____

Chapter Test continued

_____ 7. That organisms produce more offspring than their environment can support and that they compete with one another to survive are
   a. elements of natural selection.
   b. not elements of evolution.
   c. the only mechanisms of evolution.
   d. the beginning of speciation.

_____ 8. Natural selection is the process by which
   a. the age of Earth is calculated.
   b. organisms with traits well suited to the environment survive and reproduce at a greater rate than other organisms.
   c. acquired traits are passed from one generation to the next.
   d. All of the above

_____ 9. The theory of evolution predicts that
   a. closely related species will show similarities in nucleotide sequences.
   b. if species have changed over time, their genes should have changed.
   c. closely related species will show similarities in amino acid sequences.
   d. All of the above

Questions 10–12 refer to the figures below.

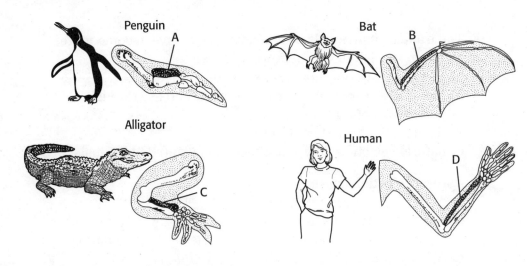

_____ 10. The bones labeled *A–D* are known as
   a. vestigial structures.   c. homologous structures.
   b. divergent structures.   d. embryonic structures.

_____ 11. The similarity of these structures suggests that the organisms
   a. have a common ancestor.   c. evolved slowly.
   b. all grow at different rates.   d. live for a long time.

Chapter Test *continued*

_____ 12. An analysis of the DNA from these organisms would indicate that
   a. their DNA is identical.
   b. they all have pharyngeal pouches.
   c. their nucleotide sequences show many similarities.
   d. they all have the same number of chromosomes.

_____ 13. Strong evidence for evolution comes from
   a. forensic biology.
   b. phylogenetic trees.
   c. works of philosophy.
   d. the fossil record

_____ 14. Structures that are present in an organism but are reduced in size and have little or no function are called
   a. pharyngeal pouches.
   b. vestigial structures.
   c. convergent structures.
   d. embryological homologies.

_____ 15. Darwin theorized that natural selection is
   a. the mechanism of evolution.
   b. how modern species have come to exist.
   c. the explanation for beak variation in finches.
   d. All of the above

In the space provided, write the letter of the description that best matches the term or phrase.

_____ 16. evolution

_____ 17. population increase

_____ 18. natural selection

_____ 19. gradualism

_____ 20. punctuated equilibrium

a. The number of individuals who possess favorable characteristics will increase in a population.

b. Species change in spurts with vast amounts of time between changes.

c. Species change over time—descent with modification.

d. Human populations do not grow unchecked because of disease, war, and famine.

e. Species change slowly and constantly.

Name _____ Class _____ Date _____

Assessment

Chapter Test

The Theory of Evolution

In the space provided, write the letter of the term or phrase that best completes each statement or best answers each question.

_____ 1. The part of Lamarck's hypothesis of evolution that proved to be correct was that
   a. evolution is linked to an organism's environmental conditions.
   b. evolution relies on the use and disuse of physical features.
   c. acquired traits are passed on to offspring.
   d. acquired traits develop slowly and gradually.

_____ 2. The forelimbs of vertebrates
   a. serve the same function.
   b. contain the same kinds of bones.
   c. have different kinds of bones.
   d. evolved from wings.

_____ 3. There is clear evidence from fossils and other sources that the species now on Earth
   a. have evolved from ancestral species that are extinct.
   b. have reached the ultimate in evolution.
   c. will eventually evolve into the same species.
   d. have always been the way they are now.

_____ 4. In South America, Darwin found fossils of armadillos that were
   a. identical to the armadillos living there.
   b. similar to the armadillos living there.
   c. identical to the armadillos living on the Galápagos Islands.
   d. similar to the armadillos living on the Galápagos Islands.

_____ 5. Darwin developed the theory of evolution by natural selection from
   a. Malthus's principles of population.
   b. the observations he made during the voyage of the *Beagle*.
   c. his experience in breeding domestic animals.
   d. All of the above

_____ 6. Subspecies can interbreed, but they
   a. have separate reproductive strategies.
   b. are separate species.
   c. are genetically different.
   d. have a very different appearance.

Name _____ Class _____ Date _____

Chapter Test *continued*

_____ 7. Fossils have been found that provide a link in the evolution of whales from
   a. two-legged mammals.
   b. legless amphibians.
   c. four-legged animals.
   d. flightless birds.

_____ 8. Species may have changed over time, and the genes that determine those species
   a. are less complex.
   b. are not important in their evolution.
   c. have not changed.
   d. have changed as well.

_____ 9. The fossil record seems to provide evidence for
   a. intermediate forms.
   b. punctuated equilibrium.
   c. natural selection.
   d. All of the above

_____ 10. The characteristics of the individuals best suited to a particular environment tend to
   a. increase in a population over time.
   b. decrease in a population over time.
   c. stay the same.
   d. fluctuate according to the weather.

_____ 11. The final step of speciation is
   a. reproductive isolation.
   b. divergence.
   c. geographic isolation.
   d. subspeciation.

_____ 12. One mechanism for reproductive isolation is
   a. genetic isolation.
   b. successful mating.
   c. geographic isolation.
   d. extinction.

_____ 13. Which statement best defines the concept of genetic variation?
   a. Offspring that do not survive do not pass their genes on to future generations.
   b. In any population, there is an array of individuals that differ slightly from one another.
   c. Individuals that are better able to cope with environmental conditions leave more offspring.
   d. The environment dictates the amount and direction of change.

_____ 14. An example of natural selection in an observable time frame is
   a. the development of an aquatic lifestyle in whales.
   b. proliferation of intermediate species between fishes and amphibians.
   c. antibiotic resistance in bacteria.
   d. the evolution of multicellular life-forms.

Name _____ Class _____ Date _____

Chapter Test continued

_____ 15. On the Galápagos Islands, the numbers of finches with different beak shapes are changed by natural selection in response to the
   a. available food supply.
   b. presence of humans.
   c. number of predators.
   d. competition among species.

In the space provided, write the letter of the description that best matches the term or phrase.

_____ 16. adaptation

_____ 17. reproductive isolation

_____ 18. punctuated equilibrium

_____ 19. homologous structures

_____ 20. tuberculosis

_____ 21. paleontologist

_____ 22. mutations

a. periods of rapid change in species separated by long periods of little or no change

b. similar anatomical arrangements of body parts of organisms indicating a shared common ancestor

c. uses radiometric dating to determine the age of fossils

d. a disease becoming difficult to treat because of the evolution of antibiotic resistance in the infecting bacteria

e. a change in DNA that is one source of new variations for natural selection to act upon

f. a feature that has become common in a population because it provides a selective advantage

g. the condition in which two populations of the same species do not breed with one another because of their geographic separation

Chapter Test continued

Read each question, and write your answer in the space provided.

23. What role does the environment play in natural selection?

24. Explain what comparing a human hemoglobin protein with the same hemoglobin protein of other species can tell about evolution.

25. Explain what Darwin meant when he referred to "the preservation of favorable variations" as natural selection.

Name _____ Class _____ Date _____

Quick Lab

DATASHEET FOR IN-TEXT LAB

Modeling Natural Selection

By making a simple model of natural selection, you can begin to understand how natural selection changes a population.

MATERIALS
- paper
- pencil
- watch or stopwatch

Procedure

1. You will be using the data table provided to record your data.

2. Write each of the following words on separate pieces of paper: *live, die, reproduce, mutate*. Fold each piece of paper in half twice so that you cannot see the words. Shuffle your folded pieces of paper.

3. Exchange two of your pieces of paper with those of a classmate. Make as many exchanges with additional classmates as you can in 30 seconds. Mix your pieces of paper between each exchange you make.

4. Look at your pieces of paper. If you have two pieces that say "die" or two pieces that say "mutate," then sit down. If you do not, then you are a "survivor." Record your results in the data table.

Data Table			
Student name	Trial 1	Trial 2	Trial 3

Name _____ Class _____ Date _____

Modeling Natural Selection continued

5. If you are a "survivor," record the words you are holding in the data table. Then refold your pieces of paper and repeat steps 2 and 3 two more times with other "survivors."

Analysis

1. **Identify** what the four slips of paper represent.

   _____

   _____

   _____

2. **Describe** what happens to most mutations in this model.

   _____

   _____

   _____

3. **Identify** what factor(s) determined who "survived." Explain.

   _____

   _____

   _____

4. **Evaluate** the shortcomings of this model of natural selection.

   _____

   _____

   _____

   _____

Name _____ Class _____ Date _____

Math Lab
Analyzing Change in Lizard Populations

DATASHEET FOR IN-TEXT LAB

Background

In 1991, Jonathan Losos, an American scientist, measured hind-limb length of lizards from several islands and the average perch diameter of the island plants. The lizards were descended from a common population 20 years earlier, and the islands had different kinds of plants on which the lizards perched. Examine the graph and answer the questions that follow.

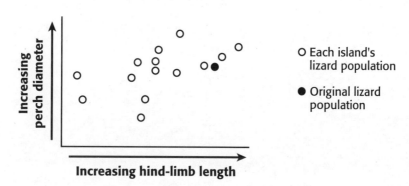

Analysis

1. **Interpreting Graphics** How did the average hind-limb length of each island's lizard population change from that of the original population?

   _____

   _____

2. **Predict** what would happen to a population of lizards with short hind limbs if they were placed on an island with a larger average perch diameter than from where they came.

   _____

   _____

3. **Justify** the argument that this experiment supports the theory of evolution by natural selection.

   _____

   _____

   _____

   _____

Name _____ Class _____ Date _____

Exploration Lab

DATASHEET FOR IN-TEXT LAB

Modeling Natural Selection

SKILLS
- Modeling a process
- Inferring relationships

OBJECTIVES
- **Model** the process of selection.
- **Relate** favorable mutations to selection and evolution.

MATERIALS
- scissors
- construction paper
- cellophane tape
- soda straws
- felt-tip marker
- meterstick or tape measure
- penny or other coin
- six-sided die

Before You Begin

Natural selection occurs when organisms that have certain **traits** survive to reproduce more than organisms that lack those traits do. A population evolves when individuals with different **genotypes** survive or reproduce at different rates. In this lab, you will model the selection of favorable traits in a new generation by using a paper model of a bird—the fictitious Egyptian origami bird (*Avis papyrus*), which lives in dry regions of North Africa. Assume that only birds that can successfully fly the long distances between water sources will live long enough to breed successfully.

1. Write a definition for each boldface term in the preceding paragraph. Use a separate sheet of paper.
2. You will be using the data table provided to record your data.
3. Based on the objectives for this lab, write a question you would like to explore about the process of selection.

_____

_____

Modeling Natural Selection *continued*

Procedure

PART A: PARENTAL GENERATION

1. Cut two strips of paper, 2 × 20 cm each. Make a loop with one strip of paper, letting the paper overlap by 1 cm, and tape the loop closed. Repeat for the other strip.

2. Tape one loop 3 cm from each end of the straw, as shown. Mark the front end of the bird with a felt-tip marker. This bird represents the parental generation.

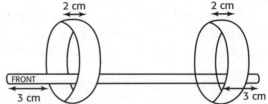

3. Test how far your parent bird can fly by releasing it with a gentle overhand pitch. Test the bird twice. Record the bird's average flight distance in the data table on the next page.

PART B: FIRST (F_1) GENERATION

4. Each origami bird lays a clutch of three eggs. Assume that one of the chicks is a clone of the parent. Use the parent to represent this chick in step 6.

5. Make two more chicks. Assume that these chicks have mutations. Follow Steps A–C below for each chick to determine the effects of its mutation.

   Step A Flip a coin to determine which end is affected by a mutation.

   Heads = anterior (front)

   Tails = posterior (back)

   Step B Throw a die to determine how the mutation affects the wing.

   - ⚀ = Wing position moves 1 cm toward the end of the straw.
   - ⚁ = Wing position moves 1 cm toward the middle of the straw.
   - ⚂ = Wing circumference increases by 2 cm.
   - ⚃ = Wing circumference decreases by 2 cm.
   - ⚄ = Wing width increases by 1 cm.
   - ⚅ = Wing width decreases by 1 cm.

   Step C A mutation is lethal if it causes a wing to fall off the straw or a wing with a circumference smaller than that of the straw. If you get a lethal mutation, disregard it and produce another chick.

6. Record the mutations and the wing dimensions of each offspring.

7. Test each bird twice by releasing it with a gentle overhand pitch. Release the birds as uniformly as possible. Record the distance each bird flies. The most successful bird is the one that flies the farthest.

Name _____ Class _____ Date _____

Modeling Natural Selection continued

Data Table

Bird	Coin flip (H or T)	Die throw (1-6)	Anterior wing (cm)			Posterior wing (cm)			Average distance flown (m)
			Width	Circum.	Distance from front	Width	Circum.	Distance from back	
Parent	NA	NA	2	19	3	2	19	3	
Generation 1									
Chick 1									
Chick 2									
Chick 3									
Generation 2									
Chick 1									
Chick 2									
Chick 3									
Generation 3									
Chick 1									
Chick 2									
Chick 3									
Generation 4									
Chick 1									
Chick 2									
Chick 3									
Generation 5									
Chick 1									
Chick 2									
Chick 3									
Generation 6									
Chick 1									
Chick 2									
Chick 3									
Generation 7									
Chick 1									
Chick 2									
Chick 3									
Generation 8									
Chick 1									
Chick 2									
Chick 3									
Generation 9									
Chick 1									
Chick 2									
Chick 3									

Copyright © by Holt, Rinehart and Winston. All rights reserved.

Holt Biology — The Theory of Evolution

Name _____ Class _____ Date _____

Modeling Natural Selection continued

PART C: SUBSEQUENT GENERATIONS

8. Assume that the most successful bird in the previous generation is the sole parent of the next generation. Repeat steps 4–7 using this bird.

9. Continue to breed, test, and record data for eight more generations.

PART D: CLEANUP AND DISPOSAL

10. Dispose of paper scraps in the designated waste container.

11. Clean up your work area and all lab equipment. Return lab equipment to its proper place. Wash your hands thoroughly before you leave the lab and after you finish all work.

Analyze and Conclude

1. Analyzing Results Did the birds you made by modeling natural selection fly farther than the first bird you made?

_____

_____

2. Inferring Conclusions How might this lab help explain the variety of species of Galápagos finches?

_____

_____

_____

_____

3. Further Inquiry Write another question about natural selection that could be explored with another investigation.

_____

_____

_____

_____

Name _____ Class _____ Date _____

Quick Lab   OBSERVATION
Comparing Limb Structure and Function

Could you tell if two people were related just by looking at them? What kinds of evidence would help you determine their relationship? In this lab, you will observe parts of various animals and look for evidence that these animals are related to one another.

OBJECTIVES

Observe and **describe** the limb structures of different organisms.

Identify relationships between the structures of different organisms.

MATERIALS

- pen or pencil

Procedure

1. Observe the forelimbs of the animals shown in **Figure 1.** Count the approximate number of bones in each of the upper and lower limbs. Record this data in **Table 1.** Then record the function of each limb.

FIGURE 1 LIMBS OF DIFFERENT ANIMALS

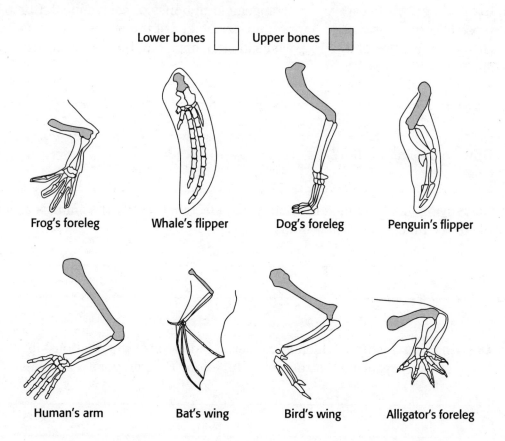

Lower bones ☐   Upper bones ■

Frog's foreleg   Whale's flipper   Dog's foreleg   Penguin's flipper

Human's arm   Bat's wing   Bird's wing   Alligator's foreleg

Copyright © by Holt, Rinehart and Winston. All rights reserved.
Holt Biology   The Theory of Evolution

Comparing Limb Structure and Function *continued*

TABLE 1 COMPARING ANIMAL LIMBS

	Approximate number of bones in upper limb	Approximate number of bones in lower limb	Function of limb
Frog			
Whale			
Dog			
Penguin			
Human			
Bat			
Bird			
Alligator			

Analysis and Conclusions

1. **Examining Data** Observe the arrangement of bones of each animal. Compare these observations with the approximate number of the bones of each animal. How are the limbs of the frog, whale, dog, human, bat, bird, and alligator similar?

   _____

   _____

   How do the limbs differ?

   _____

2. **Classifying** Look again at the data you collected. Classify the animals according to the functions of their limbs.

   _____

   _____

   _____

3. **Drawing Conclusions** Which is the better indicator of the relationship between two organisms—structure or function? Explain your reasoning.

   _____

   _____

Name _____ Class _____ Date _____

Skills Practice Lab

Melanism in Insects

MODELING

Natural selection, the reproductive success of organisms best suited to their environment, is a driving force in evolution. Natural selection occurs within *populations*, which are interbreeding groups of individuals of the same species. *Genetic variation*, the alternative types of genes for inherited traits, is one factor in the reproductive success of certain members of a population. The result of natural selection is *adaptation*, the changing of a population in a way that makes it better suited to its environment.

Industrial melanism is the term used to describe the adaptation of a population by the darkening of its individuals in response to industrial pollution. Consider a population of beetles that live on tree trunks. In the absence of pollution, the trunks of trees where these beetles live are light grayish green due to the presence of lichens. The beetles living on these trees are also light-colored and easily camouflaged on the tree trunks. Over time, however, the tree trunks become covered with soot and turn dark. Within a few decades, a dark variety of the beetle becomes more common than the light-colored variety in response to the pollution.

In this lab, you will simulate how successfully predators locate prey in different environments. Then you will relate changes in a population of beetles with two color variations to changes in the environment.

OBJECTIVES

Describe the importance of coloration in avoiding predation.

Relate environmental change to changes in organisms within an ecosystem.

Explain how natural selection causes populations to change.

MATERIALS

- colored pencils (2)
- forceps
- newspaper dots (60)
- sheet of newspaper
- sheet of white paper
- watch or clock with second hand
- white paper dots (60)

Procedure

PART 1: SIMULATING PREDATOR-PREY RELATIONSHIPS

1. Work with a partner, and decide which of you will be the "predator" and which will be the timekeeper.

2. Place a sheet of white paper on your lab table. If you are the timekeeper, scatter 30 white paper dots and 30 newspaper dots on the paper while your partner looks away. The dots represent a bird's prey. If you are the predator, use forceps to pick up as many dots as possible in 15 seconds while your lab partner watches the time. The forceps simulate a bird's beak.

Copyright © by Holt, Rinehart and Winston. All rights reserved.

Holt Biology 49 The Theory of Evolution

Melanism in Insects *continued*

3. Count the number of each type of dots picked up in 15 seconds. Record these numbers in **Table 1**.

TABLE 1 NUMBER OF PAPER DOTS SCATTERED AND RECOVERED

Trial	Background	Total number of dots scattered		Total number of dots picked up		Percent of available prey recovered	
		Newspaper	White	Contrasting background	Matching background	Contrasting background	Matching background
1							
2							
3							
4							

- How does the number of each type of dot captured compare with the number of each type of dot remaining on the paper?

_____

_____

_____

4. Replace the white paper with a sheet of newspaper. If you are the timekeeper, scatter 30 white paper dots and 30 newspaper dots on the newspaper. If you are the predator, repeat the hunting procedure while your partner watches the time. Again, record the numbers in **Table 1**.

- How does the number of each type of dot captured compare with the number of each type remaining on the newspaper?

_____

_____

_____

5. Change roles with your partner, and repeat steps 2–4.

PART 2: ANALYZING PREDATOR-PREY RELATIONSHIPS

6. Examine **Table 2**, which represents data from a 10-year study of a population of beetles native to the United States. The numbers represent beetles captured in traps that were located in the same area each year.

Melanism in Insects *continued*

TABLE 2 LIGHT BEETLES AND DARK BEETLES CAPTURED

Year	Number of light beetles captured	Number of dark beetles captured
1	710	99
2	590	122
3	502	205
4	405	215
5	255	295
6	225	357
7	202	415
8	151	499
9	85	600
10	59	730

7. Use the data in **Table 2** to construct a graph comparing the number of light beetles captured with the number of dark beetles captured. Construct your graph in **Figure 1**. Use a different colored pencil to differentiate the two forms of beetles. Label the graph curves clearly or make a key.

8. Dispose of your materials according to your teacher's directions.

9. Clean up your work area, and wash your hands before leaving the lab.

FIGURE 1 CHANGE IN COLOR IN A BEETLE POPULATION

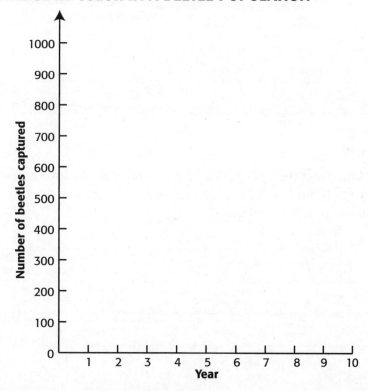

Name _____ Class _____ Date _____

Melanism in Insects *continued*

Analysis

1. **Analyzing Results** Assume that coloration is not important to successful predation. If you were a predator selecting from a field of an equal number of light prey and dark prey, you would expect to capture an approximately equal number of each color of prey. What did the experiment you conducted in Part 1 indicate?

   _____

   _____

   _____

   _____

2. **Analyzing Graphs** Using the graph you made in **Figure 1,** describe what happened in the population of beetles in the sampled area.

   _____

   _____

   _____

3. **Explaining Events** How is industrial melanism in a population of insects different from some students in your class dying their hair?

   _____

   _____

   _____

   _____

Conclusions

1. **Drawing Conclusions** From your graph, what conclusions can you make about how genes and evolutionary fitness may have contributed to the changes in the beetle population?

   _____

   _____

   _____

   _____

Melanism in Insects *continued*

2. Making Predictions What effect do you think using cleaner-burning fuels might have on a population of dark-colored insects that live on soot-covered tree trunks?

_____

_____

3. Making Predictions Assume that an increase in the dark variety of a population of beetles is an adaptive response to the darkening of tree trunks as a result of pollution. Then, over time, the pollution is reduced and the tree trunks return to their former light color. What would you expect to happen in the beetle population if the tree trunks on which they live became light again?

_____

_____

_____

_____

4. Applying Conclusions In the 1940s, DDT was used effectively as an insecticide against mosquitoes. Twenty years after the widespread use of DDT, a large proportion of mosquitoes was resistant to the insecticide. How is the rise of DDT-resistant mosquitoes similar to industrial melanism in beetles living on soot-covered trees?

_____

_____

_____

_____

_____

Extensions

1. **Research and Communications** Use the library or an on-line database to discover other organisms that have shown adaptation by industrial melanism over a short period of time.

2. **Research and Communications** The term *artificial selection* is used to describe the process by which humans change domesticated animals and plants by breeding individuals with desirable characteristics. Use the library or the Internet to discover how the use of pesticides and antibiotics has affected insects and bacteria. Write a paragraph in which you state whether the changes in populations of these organisms can be attributed to natural selection or artificial selection. Be sure to justify your position.

Name _____ Class _____ Date _____

Quick Lab

DATASHEET FOR IN-TEXT LAB

Modeling Natural Selection

By making a simple model of natural selection, you can begin to understand how natural selection changes a population.

MATERIALS
- paper
- pencil
- watch or stopwatch

Procedure

1. You will be using the data table provided to record your data.
2. Write each of the following words on separate pieces of paper: *live, die, reproduce, mutate*. Fold each piece of paper in half twice so that you cannot see the words. Shuffle your folded pieces of paper.
3. Exchange two of your pieces of paper with those of a classmate. Make as many exchanges with additional classmates as you can in 30 seconds. Mix your pieces of paper between each exchange you make.
4. Look at your pieces of paper. If you have two pieces that say "die" or two pieces that say "mutate," then sit down. If you do not, then you are a "survivor." Record your results in the data table.

Data Table

Student name	Trial 1	Trial 2	Trial 3

Copyright © by Holt, Rinehart and Winston. All rights reserved.

Holt Biology — The Theory of Evolution

TEACHER RESOURCE PAGE

Name _____ Class _____ Date _____

Modeling Natural Selection *continued*

5. If you are a "survivor," record the words you are holding in the data table. Then refold your pieces of paper and repeat steps 2 and 3 two more times with other "survivors."

Analysis

1. **Identify** what the four slips of paper represent.

   the different things that can happen to an organism if it

   is exposed to change in its environment

2. **Describe** what happens to most mutations in this model.

   Most mutations will be passed on because they are harmful only if an

   individual has two copies.

3. **Identify** what factor(s) determined who "survived." Explain.

   The survivors avoided chance death (a "die" card) and two copies of the

   mutation, which is "lethal" when expressed.

4. **Evaluate** the shortcomings of this model of natural selection.

   sample answer: does not distinguish between beneficial and harmful

   mutations; does not distinguish between living and reproducing

TEACHER RESOURCE PAGE

Name _____ Class _____ Date _____

Math Lab   **DATASHEET FOR IN-TEXT LAB**

Analyzing Change in Lizard Populations

Background

In 1991, Jonathan Losos, an American scientist, measured hind-limb length of lizards from several islands and the average perch diameter of the island plants. The lizards were descended from a common population 20 years earlier, and the islands had different kinds of plants on which the lizards perched. Examine the graph and answer the questions that follow.

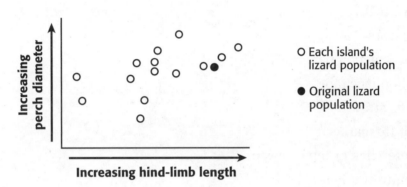

Analysis

1. **Interpreting Graphics** How did the average hind-limb length of each island's lizard population change from that of the original population?

   The average hind limb length of each population changed in response to

   differences in the average perch diameter of plants on the different islands.

2. **Predict** what would happen to a population of lizards with short hind limbs if they were placed on an island with a larger average perch diameter than from where they came.

   The population could evolve and have longer average hind limbs, or it could

   go extinct.

3. **Justify** the argument that this experiment supports the theory of evolution by natural selection.

   The experiment illustrates that characteristics of populations can change

   over time in response to environmental pressures.

TEACHER RESOURCE PAGE

Name _____ Class _____ Date _____

Exploration Lab

DATASHEET FOR IN-TEXT LAB

Modeling Natural Selection

SKILLS
- Modeling a process
- Inferring relationships

OBJECTIVES
- **Model** the process of selection.
- **Relate** favorable mutations to selection and evolution.

MATERIALS
- scissors
- construction paper
- cellophane tape
- soda straws
- felt-tip marker
- meterstick or tape measure
- penny or other coin
- six-sided die

Before You Begin

Natural selection occurs when organisms that have certain **traits** survive to reproduce more than organisms that lack those traits do. A population evolves when individuals with different **genotypes** survive or reproduce at different rates. In this lab, you will model the selection of favorable traits in a new generation by using a paper model of a bird—the fictitious Egyptian origami bird (*Avis papyrus*), which lives in dry regions of North Africa. Assume that only birds that can successfully fly the long distances between water sources will live long enough to breed successfully.

1. Write a definition for each boldface term in the preceding paragraph. Use a separate sheet of paper. **Answers appear in the TE for this lab.**

2. You will be using the data table provided to record your data.

3. Based on the objectives for this lab, write a question you would like to explore about the process of selection.

   Answers will vary. For example: Are the effects of natural selection obvious

   in only a few generations?

Copyright © by Holt, Rinehart and Winston. All rights reserved.
Holt Biology — The Theory of Evolution

Modeling Natural Selection *continued*

Procedure

PART A: PARENTAL GENERATION

1. Cut two strips of paper, 2 × 20 cm each. Make a loop with one strip of paper, letting the paper overlap by 1 cm, and tape the loop closed. Repeat for the other strip.

2. Tape one loop 3 cm from each end of the straw, as shown. Mark the front end of the bird with a felt-tip marker. This bird represents the parental generation.

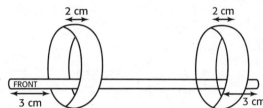

3. Test how far your parent bird can fly by releasing it with a gentle overhand pitch. Test the bird twice. Record the bird's average flight distance in the data table on the next page.

PART B: FIRST (F_1) GENERATION

4. Each origami bird lays a clutch of three eggs. Assume that one of the chicks is a clone of the parent. Use the parent to represent this chick in step 6.

5. Make two more chicks. Assume that these chicks have mutations. Follow Steps A–C below for each chick to determine the effects of its mutation.

   Step A Flip a coin to determine which end is affected by a mutation.

   Heads = anterior (front)

   Tails = posterior (back)

   Step B Throw a die to determine how the mutation affects the wing.

   ⚀ = Wing position moves 1 cm toward the end of the straw.

   ⚁ = Wing position moves 1 cm toward the middle of the straw.

   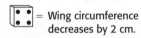 = Wing circumference increases by 2 cm.

   ⚄ = Wing circumference decreases by 2 cm.

   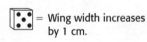 = Wing width increases by 1 cm.

   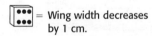 = Wing width decreases by 1 cm.

   Step C A mutation is lethal if it causes a wing to fall off the straw or a wing with a circumference smaller than that of the straw. If you get a lethal mutation, disregard it and produce another chick.

6. Record the mutations and the wing dimensions of each offspring.

7. Test each bird twice by releasing it with a gentle overhand pitch. Release the birds as uniformly as possible. Record the distance each bird flies. The most successful bird is the one that flies the farthest.

TEACHER RESOURCE PAGE

Name _____ Class _____ Date _____

Modeling Natural Selection continued

Data Table

Bird	Coin flip (H or T)	Die throw (1–6)	Anterior wing (cm)			Posterior wing (cm)			Average distance flown (m)
			Width	Circum.	Distance from front	Width	Circum.	Distance from back	
Parent	NA	NA	2	19	3	2	19	3	
Generation 1									
Chick 1									
Chick 2									
Chick 3									
Generation 2									
Chick 1									
Chick 2									
Chick 3									
Generation 3									
Chick 1									
Chick 2									
Chick 3									
Generation 4									
Chick 1									
Chick 2									
Chick 3									
Generation 5									
Chick 1									
Chick 2									
Chick 3									
Generation 6									
Chick 1									
Chick 2									
Chick 3									
Generation 7									
Chick 1									
Chick 2									
Chick 3									
Generation 8									
Chick 1									
Chick 2									
Chick 3									
Generation 9									
Chick 1									
Chick 2									
Chick 3									

Copyright © by Holt, Rinehart and Winston. All rights reserved.

Holt Biology — The Theory of Evolution

Modeling Natural Selection *continued*

PART C: SUBSEQUENT GENERATIONS

8. Assume that the most successful bird in the previous generation is the sole parent of the next generation. Repeat steps 4–7 using this bird.
9. Continue to breed, test, and record data for eight more generations.

PART D: CLEANUP AND DISPOSAL

10. Dispose of paper scraps in the designated waste container.
11. Clean up your work area and all lab equipment. Return lab equipment to its proper place. Wash your hands thoroughly before you leave the lab and after you finish all work.

Analyze and Conclude

1. **Analyzing Results** Did the birds you made by modeling natural selection fly farther than the first bird you made?

   Most students should answer "yes" as the best-flying birds are selected as the sole parents of the next generation.

2. **Inferring Conclusions** How might this lab help explain the variety of species of Galápagos finches?

   This lab demonstrates that organisms can change significantly over only a few generations. The lab therefore shows how isolated populations could diverge to the point that they constitute different species, as happened to finches on the Galápagos Islands.

3. **Further Inquiry** Write another question about natural selection that could be explored with another investigation.

   Answers will vary. For example: Does natural selection act on one trait at a time, or can selection pressure affect the evolution of several traits at once?

TEACHER RESOURCE PAGE

Quick Lab   OBSERVATION
Comparing Limb Structure and Function

Teacher Notes

TIME REQUIRED 20 minutes

SKILLS ACQUIRED
Classifying
Identifying patterns
Inferring
Interpreting
Organizing and analyzing data

RATINGS Easy ←1——2——3——4→ Hard
Teacher Prep–1
Student Setup–1
Concept Level–1
Cleanup–1

THE SCIENTIFIC METHOD

Make Observations Students observe the bones in the limbs of various animals and look for evidence that the animals are related to one another.

Analyze the Results Students analyze their data in Analysis and Conclusions questions 1 and 2.

Draw Conclusions Analysis and Conclusions question 3 asks student to draw conclusions from their data.

MATERIALS
Skeletons of vertebrate limbs would enhance this lab.

TIPS AND TRICKS

This lab works best in groups of two but can be done individually.

Stress that biologists often must rely on observations made from the drawings of others, at least during the initial stages of an investigation. Explain that in this lab, students must make the best use of the drawings of appendages.

Show students the key in Figure 1, which distinguishes upper bones from lower bones. Make sure students are able to compare the humerus (upper bone) and radius and ulna (lower bones) among the animals in the diagrams.

Allow students to consult other sources for more detailed diagrams of the upper and lower limbs of animals.

The bone counts will be approximations based on what students can see and count in the drawings in Figure 1. The data in Table 1 also are based on the drawings, not necessarily the actual number of bones in those vertebrates.

Copyright © by Holt, Rinehart and Winston. All rights reserved.

Holt Biology   The Theory of Evolution

TEACHER RESOURCE PAGE

Name _____ Class _____ Date _____

Quick Lab   OBSERVATION
Comparing Limb Structure and Function

Could you tell if two people were related just by looking at them? What kinds of evidence would help you determine their relationship? In this lab, you will observe parts of various animals and look for evidence that these animals are related to one another.

OBJECTIVES

Observe and **describe** the limb structures of different organisms.

Identify relationships between the structures of different organisms.

MATERIALS

- pen or pencil

Procedure

1. Observe the forelimbs of the animals shown in **Figure 1**. Count the approximate number of bones in each of the upper and lower limbs. Record this data in **Table 1**. Then record the function of each limb.

FIGURE 1 LIMBS OF DIFFERENT ANIMALS

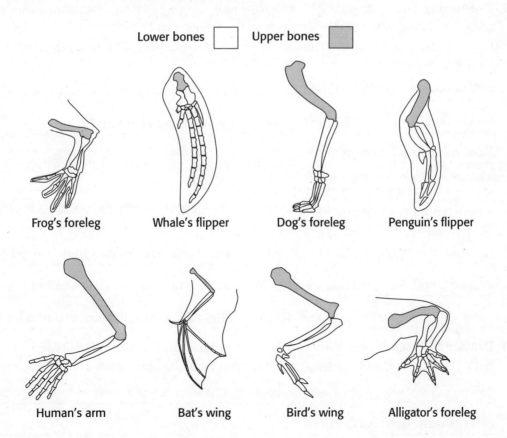

Lower bones ☐   Upper bones ▓

Frog's foreleg   Whale's flipper   Dog's foreleg   Penguin's flipper

Human's arm   Bat's wing   Bird's wing   Alligator's foreleg

Copyright © by Holt, Rinehart and Winston. All rights reserved.

Holt Biology   The Theory of Evolution

TEACHER RESOURCE PAGE

Name _____ Class _____ Date _____

Comparing Limb Structure and Function *continued*

TABLE 1 COMPARING ANIMAL LIMBS

	Approximate number of bones in upper limb	Approximate number of bones in lower limb	Function of limb
Frog	1	15	locomotion on land
Whale	1	31	swimming
Dog	1	23	locomotion on land
Penguin	1	9	swimming
Human	1	28	grasping and manipulating
Bat	1	26	flight
Bird	1	9	flight
Alligator	1	28	locomotion on land

Analysis and Conclusions

1. **Examining Data** Observe the arrangement of bones of each animal. Compare these observations with the approximate number of the bones of each animal. How are the limbs of the frog, whale, dog, human, bat, bird, and alligator similar?

   Answers may vary but should indicate that the number and arrangement of

   bones in the upper and lower limbs of the animals are similar.

   How do the limbs differ?

   The shape and function of the limbs may differ.

2. **Classifying** Look again at the data you collected. Classify the animals according to the functions of their limbs.

   Appendages of the bird and bat are used for flight. The human limb is used for

   grasping and manipulating, the whale's and penguin's limbs are used in swim-

   ming, and the frog's, dog's, and alligator's limbs are used in locomotion on land.

3. **Drawing Conclusions** Which is the better indicator of the relationship between two organisms—structure or function? Explain your reasoning.

   Answers may vary but should suggest that structure is a better indicator of

   relationship than is function.

TEACHER RESOURCE PAGE

Skills Practice Lab

Melanism in Insects

MODELING

Teacher Notes

TIME REQUIRED One 45-minute period

SKILLS ACQUIRED
Collecting data
Identifying patterns
Inferring
Interpreting
Organizing and analyzing data

RATINGS
Teacher Prep–2
Student Setup–2
Concept Level–2
Cleanup–2

THE SCIENTIFIC METHOD

Make Observations Students observe a model of adaptation.

Analyze the Results Analysis questions 1 and 2 require students to analyze the results of their model.

Draw Conclusions Students draw conclusions in Conclusions question 1.

Communicate the Results Step 7 of the Procedure asks student to graphically communicate the results of a study.

MATERIALS

Materials for this lab can be purchased from WARD'S. See the *Master Materials List* for ordering instructions.

TIPS AND TRICKS
Preparation

This lab works best in groups of two students.

The newspaper dots and white paper dots should be punched out, using a hole punch, and counted ahead of time. They can be stored in plastic bags for each group. You can laminate the paper you use to make the dots if you want to keep them for future labs. The sheets of white paper and newspaper can be laminated to avoid tearing while being used.

Ideally, the newspaper in step 4 should be the same size as the white paper in step 2. Try to avoid using sheets of newspaper with full-color images, both for the dots or for the base.

TEACHER RESOURCE PAGE

Name _____ Class _____ Date _____

Skills Practice Lab

MODELING

Melanism in Insects

Natural selection, the reproductive success of organisms best suited to their environment, is a driving force in evolution. Natural selection occurs within *populations*, which are interbreeding groups of individuals of the same species. *Genetic variation*, the alternative types of genes for inherited traits, is one factor in the reproductive success of certain members of a population. The result of natural selection is *adaptation*, the changing of a population in a way that makes it better suited to its environment.

Industrial melanism is the term used to describe the adaptation of a population by the darkening of its individuals in response to industrial pollution. Consider a population of beetles that live on tree trunks. In the absence of pollution, the trunks of trees where these beetles live are light grayish green due to the presence of lichens. The beetles living on these trees are also light-colored and easily camouflaged on the tree trunks. Over time, however, the tree trunks become covered with soot and turn dark. Within a few decades, a dark variety of the beetle becomes more common than the light-colored variety in response to the pollution.

In this lab, you will simulate how successfully predators locate prey in different environments. Then you will relate changes in a population of beetles with two color variations to changes in the environment.

OBJECTIVES

Describe the importance of coloration in avoiding predation.

Relate environmental change to changes in organisms within an ecosystem.

Explain how natural selection causes populations to change.

MATERIALS

- colored pencils (2)
- forceps
- newspaper dots (60)
- sheet of newspaper
- sheet of white paper
- watch or clock with second hand
- white paper dots (60)

Procedure

PART 1: SIMULATING PREDATOR-PREY RELATIONSHIPS

1. Work with a partner, and decide which of you will be the "predator" and which will be the timekeeper.

2. Place a sheet of white paper on your lab table. If you are the timekeeper, scatter 30 white paper dots and 30 newspaper dots on the paper while your partner looks away. The dots represent a bird's prey. If you are the predator, use forceps to pick up as many dots as possible in 15 seconds while your lab partner watches the time. The forceps simulate a bird's beak.

Melanism in Insects continued

3. Count the number of each type of dots picked up in 15 seconds. Record these numbers in **Table 1**.

TABLE 1 NUMBER OF PAPER DOTS SCATTERED AND RECOVERED

Trial	Total number of dots scattered			Total number of dots picked up		Percent of available prey recovered	
	Background	Newspaper	White	Contrasting background	Matching background	Contrasting background	Matching background
1							
2							
3				Answers	will vary.		
4							

- How does the number of each type of dot captured compare with the number of each type of dot remaining on the paper?

  Answer may vary, but most students should discover that more of the newspaper dots were captured, and more of the white paper dots remained on the paper.

4. Replace the white paper with a sheet of newspaper. If you are the timekeeper, scatter 30 white paper dots and 30 newspaper dots on the newspaper. If you are the predator, repeat the hunting procedure while your partner watches the time. Again, record the numbers in **Table 1**.

   - How does the number of each type of dot captured compare with the number of each type remaining on the newspaper?

     Answer may vary, but most students should discover that more of the white paper dots were recovered, and more of the newspaper dots remained on the newspaper.

5. Change roles with your partner, and repeat steps 2–4.

PART 2: ANALYZING PREDATOR-PREY RELATIONSHIPS

6. Examine **Table 2,** which represents data from a 10-year study of a population of beetles native to the United States. The numbers represent beetles captured in traps that were located in the same area each year.

TEACHER RESOURCE PAGE

Name _____ Class _____ Date _____

Melanism in Insects *continued*

TABLE 2 LIGHT BEETLES AND DARK BEETLES CAPTURED

Year	Number of light beetles captured	Number of dark beetles captured
1	710	99
2	590	122
3	502	205
4	405	215
5	255	295
6	225	357
7	202	415
8	151	499
9	85	600
10	59	730

7. Use the data in **Table 2** to construct a graph comparing the number of light beetles captured with the number of dark beetles captured. Construct your graph in **Figure 1**. Use a different colored pencil to differentiate the two forms of beetles. Label the graph curves clearly or make a key.

8. Dispose of your materials according to your teacher's directions.

9. Clean up your work area, and wash your hands before leaving the lab.

FIGURE 1 CHANGE IN COLOR IN A BEETLE POPULATION

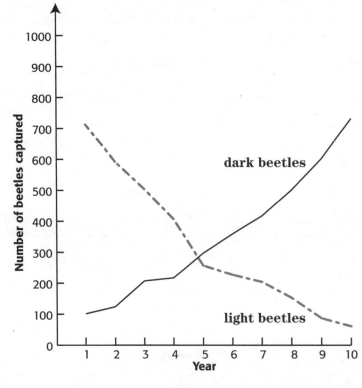

Holt Biology — The Theory of Evolution

Name _____ Class _____ Date _____

Melanism in Insects *continued*

Analysis

1. **Analyzing Results** Assume that coloration is not important to successful predation. If you were a predator selecting from a field of an equal number of light prey and dark prey, you would expect to capture an approximately equal number of each color of prey. What did the experiment you conducted in Part 1 indicate?

   Answers will vary. Most students will note that they did not capture an equal

   number of different-colored prey and that they captured a larger number of

   the prey that did not blend with the background. These results support the

   idea that coloration is important to the predator for successful predation

   and equally important to the prey to avoid predation.

2. **Analyzing Graphs** Using the graph you made in **Figure 1,** describe what happened in the population of beetles in the sampled area.

   Over the period, the number of light-colored beetles decreased and the

   number of dark-colored beetles increased.

3. **Explaining Events** How is industrial melanism in a population of insects different from some students in your class dying their hair?

   Melanism in insects is the result of genetic variation and results in a popula-

   tion being better adapted to its environment. Students dyeing their hair is

   not a reflection of genetic variation, natural selection, or evolutionary

   adaptation.

Conclusions

1. **Drawing Conclusions** From your graph, what conclusions can you make about how genes and evolutionary fitness may have contributed to the changes in the beetle population?

   Over the period, the number of beetles having a gene for light color

   decreased, and the number of beetles having a gene for dark color increased.

   The dark beetles were apparently better adapted to the environment than

   the light beetles.

Name _____ Class _____ Date _____

Melanism in Insects *continued*

2. **Making Predictions** What effect do you think using cleaner-burning fuels might have on a population of dark-colored insects that live on soot-covered tree trunks?

   The probable result would be a decrease in the number of dark-colored insects and an increase in the number of light-colored insects.

3. **Making Predictions** Assume that an increase in the dark variety of a population of beetles is an adaptive response to the darkening of tree trunks as a result of pollution. Then, over time, the pollution is reduced and the tree trunks return to their former light color. What would you expect to happen in the beetle population if the tree trunks on which they live became light again?

   Answers will vary. Students should indicate that the number of light-colored beetles would increase and the number of dark-colored beetles would decrease; or that the ratio of light-colored to dark-colored beetles would return to the level before the pollution.

4. **Applying Conclusions** In the 1940s, DDT was used effectively as an insecticide against mosquitoes. Twenty years after the widespread use of DDT, a large proportion of mosquitoes was resistant to the insecticide. How is the rise of DDT-resistant mosquitoes similar to industrial melanism in beetles living on soot-covered trees?

   Answers will vary. Students should recognize that the application of DDT, like the darkening of tree trunks with soot, resulted in the death of many individuals that were not adapted to this environmental change. However, a small number of individuals survived to reproduce. Over generations, the number of individuals having this adaptation grew.

Extensions

1. **Research and Communications** Use the library or an on-line database to discover other organisms that have shown adaptation by industrial melanism over a short period of time.

2. **Research and Communications** The term *artificial selection* is used to describe the process by which humans change domesticated animals and plants by breeding individuals with desirable characteristics. Use the library or the Internet to discover how the use of pesticides and antibiotics has affected insects and bacteria. Write a paragraph in which you state whether the changes in populations of these organisms can be attributed to natural selection or artificial selection. Be sure to justify your position.

Answer Key

Directed Reading

SECTION: THE THEORY OF EVOLUTION BY NATURAL SELECTION
1. b
2. a
3. c
4. b
5. Lamarck proposed that during an individual's lifetime, structures increase or decrease in size according to their use. These acquired traits are then passed on to the offspring.
6. A species is a group of genetically similar organisms that can interbreed and produce fertile offspring. A population is a group of organisms of the same species living in a given place at the same time.
7. Adaptation is the changing of a species to become better suited to its environment. Natural selection is the process by which populations change in response to their environment. Adaptation occurs as a result of natural selection.
8. more
9. Genes
10. Natural selection
11. Isolation
12. species

SECTION: EVIDENCE OF EVOLUTION
1. b
2. c
3. d
4. c
5. a
6. d
7. a
8. c
9. b
10. fewer
11. one
12. 67
13. eight
14. gene
15. Scientists compare the nucleotide sequence of genes.
16. If evolution has taken place, then species descended from a recent common ancestor should have fewer amino acid differences than species that share a common ancestor in the distant past.

SECTION: EXAMPLES OF EVOLUTION
1. environment
2. offspring
3. variation
4. compete
5. more
6. Some bacteria have developed a mutation that allows them to grow in the presence of the antibiotic that destroys bacteria without the mutation. The mutant bacteria become more prevalent in the population and eventually predominant.
7. The Grants measured the beaks of many Galápagos finches year after year.
8. The environmental challenge was the decrease in the number of small, tender seeds during dry years. This decrease required the birds to eat larger, tougher seeds.
9. Natural selection increased the average beak size during dry years and decreased the beak size during wet years.
10. The accumulation of differences between groups is called divergence. Speciation is the process by which new species form.
11. Subspecies are populations of the same species that differ genetically because of adaptations to different living conditions. When subspecies become so different that they cannot interbreed, they are considered separate species.
12. barrier
13. isolation
14. species

Active Reading

SECTION: THE THEORY OF EVOLUTION BY NATURAL SELECTION
1. Malthus believed that while every human has the potential to produce many offspring during his or her lifetime, only a limited number of those offspring survive to further reproduce.

2. observations made on his voyage and his experiences breeding domestic animals
3. Certain individuals have physical or behavioral traits that suit their environment. Because of these traits, the individuals are more likely to survive and reproduce offspring with the same traits. Over time, the number of individuals possessing the traits exceeds the number of individuals lacking the traits, which changes the nature of the population.
4. genes
5. As the number of individuals carrying the alleles for a certain trait increases, the frequency of that trait increases in a population.
6. mutations and recombination of alleles during sexual reproduction

SECTION: EVIDENCE OF EVOLUTION

1. Many species have lived in areas where fossils do not form.
2. Fossils are most likely to form in wet lowlands, slow-moving streams, lakes, shallow seas, and areas near volcanoes that spew out volcanic ash. Most fossils form when organisms are rapidly buried in fine sediments.
3. Many organisms decay before sediments cover them, or they are eaten and scattered by scavengers.

SECTION: EXAMPLES OF EVOLUTION

1. The title of the graph is "Beak-Size Variation." An observer would expect to discover data regarding differences in the beak size of the members of a certain population.
2. The horizontal axis identifies various years.
3. Each interval on the axis represents 1 year.
4. beak size
5. millimeters
6. Beak size increases.
7. Beak size decreases.
8. c

Vocabulary Review

1. d
2. a
3. b
4. c
5. b
6. c
7. c
8. d
9. c
10. b
11. c
12. a
13. a
14. c
15. c
16. c
17. b
18. d

Science Skills

APPLYING INFORMATION

1. Although both male and female African elephants normally have tusks, a few elephants are born without tusks because of a genetic mutation.
2. Other animals can threaten an elephant's ability to survive. In this case, humans acted as predators.
3. Competition for resources, such as food and water, was not a significant factor because poachers greatly reduced the population.
4. Elephants without tusks were not killed and therefore those with the genes that resulted in not having tusks survived to produce more offspring without tusks.
5. Each generation of elephants contained a higher proportion of individuals with the genes for being tuskless. Over time, tuskless elephants, which are safe from ivory poachers, became more common.

Concept Mapping

1. punctuated equilibrium, gradualism
2. gradualism, punctuated equilibrium
3. natural selection
4. finch beaks
5. fossils
6. antibiotic environment
7. vestigial structures
8. homologous structures
9. embryonic development
10. nucleic acids, proteins
11. proteins, nucleic acids

Critical Thinking

1. b
2. a
3. d
4. c
5. b
6. g
7. h
8. c
9. a
10. e
11. f
12. d
13. c, f
14. h, a
15. b, g
16. d, e
17. d
18. c
19. c

Test Prep Pretest

1. b
2. d
3. b
4. a
5. a
6. d
7. c
8. b
9. c
10. d
11. extinct
12. Lamarck's
13. adaptation
14. population
15. isolation
16. amino acid, nucleotide
17. homologous
18. traits
19. gradualism
20. vestigial
21. evolution, gradual or slow
22. divergence
23. Lamarck thought that evolution occurred as structures developed through use or disappeared because of lack of use. He thought that these acquired characteristics could be passed on to offspring.
24. Malthus stated that the human population had the potential to increase much faster than the food supply but did not do so because of disease, war, and famine. Darwin realized that Malthus's principles of population applied not only to humans but also to all species, and Darwin took this into consideration when he developed his theory of evolution by natural selection.
25. Scientists now know that genetics is the basis of inherited traits. Individuals with favorable traits become more common in a population because natural selection causes the frequency of certain alleles in a population to increase or decrease over time.

Quiz

SECTION: THE THEORY OF EVOLUTION BY NATURAL SELECTION

1. d
2. a
3. c
4. b
5. c
6. c
7. e
8. b
9. d
10. a

SECTION: EVIDENCE OF EVOLUTION

1. c
2. a
3. d
4. b
5. b
6. d
7. a
8. e
9. b
10. c

SECTION: EXAMPLES OF EVOLUTION

1. b
2. b
3. a
4. c
5. b
6. a
7. b
8. a
9. a
10. b

Chapter Test (General)

1. b
2. c
3. b
4. b
5. c
6. c
7. a
8. b
9. d
10. c
11. a
12. c
13. d
14. b
15. d
16. c
17. d
18. a
19. e
20. b

Chapter Test (Advanced)

1. a
2. b
3. a
4. b
5. d
6. c
7. c
8. d
9. d
10. a
11. a
12. c
13. b
14. c
15. a
16. f
17. g
18. a
19. b
20. d
21. c
22. e

23. According to Darwin's theory, those organisms with traits that are best suited to the environment are more likely to survive and successfully reproduce than those that do not have such traits.

24. Species that shared a common ancestor more recently, such as humans and gorillas, have few amino acid sequence differences. Species that shared an ancestor in the distant past, such as gorillas and frogs, will have many amino acid sequence differences.

25. Individuals having any slight advantage over others would have the best chance of surviving and reproducing. Any variation that decreases the likelihood of reproducing would be destroyed. Thus, the favorable variations would be preserved.

TEACHER RESOURCE PAGE

Lesson Plan

Section: The Theory of Evolution by Natural Selection

Pacing

Regular Schedule: with lab(s): N/A   without lab(s): 3 days
Block Schedule: with lab(s): N/A   without lab(s): 1 1/2 days

Objectives

1. Identify several observations that led Darwin to conclude that species evolve.
2. Relate the process of natural selection to its outcome.
3. Summarize the main points of Darwin's theory of evolution by natural selection as it is stated today.
4. Contrast the gradualism and puctuated equilibrium models of evolution.

National Science Education Standards Covered

UNIFYING CONCEPTS AND PROCESSES

UCP1: Systems, order, and organization

UCP2: Evidence, models, and explanation

UCP4: Evolution and equilibrium

UCP5: Form and function

SCIENCE AS INQUIRY

SI1: Abilities necessary to do scientific inquiry

SI2: Understandings about scientific inquiry

HISTORY AND NATURE OF SCIENCE

HNS1: Science as a human endeavor

HNS3: Historical perspectives

LIFE SCIENCE: MOLECULAR BASIS OF HEREDITY

LSGene3: Changes in DNA (mutations) occur spontaneously at low rates.

LIFE SCIENCE: BIOLOGICAL EVOLUTION

LSEvol1: Species evolve over time.

LSEvol2: The great diversity of organisms is the result of more than 3.5 billion years of evolution.

Copyright © by Holt, Rinehart and Winston. All rights reserved.

Holt Biology   The Theory of Evolution

TEACHER RESOURCE PAGE

Lesson Plan *continued*

LSEvol3: Natural selection and its evolutionary consequences provide a scientific explanation for the fossil record of ancient life forms, as well as for the striking molecular similarities observed among the diverse species of living organisms.

LSEvol4: The millions of different species of plants, animals, and microorganisms that live on earth today are related by descent from common ancestors.

LSEvol5: Biological classifications are based on how organisms are related.

LIFE SCIENCE: MATTER, ENERGY, AND ORGANIZATION IN LIVING SYSTEMS

LSMat4: The complexity and organization of organisms accommodates the need for obtaining, transforming, transporting, releasing, and eliminating the matter and energy used to sustain the organism.

LSMat5: The distribution and abundance of organisms and populations in ecosystems are limited by the availability of matter and energy and the ability of the ecosystem to recycle materials.

LSMat6: As matter and energy flow through different levels of organization of living systems—cells, organs, communities—and between living systems and the physical environment, chemical elements are recombined in different ways.

KEY
SE = Student Edition   TE = Teacher Edition
CRF = Chapter Resource File

Block 1

CHAPTER OPENER (*45 minutes*)

- **Quick Review,** SE. Students answer questions covered in previous sections of the textbook as preparation for the chapter content. (**GENERAL**)

- **Reading Activity,** SE. Students create a Reader Response Log to record their personal responses to the concepts presented in the chapter. (**GENERAL**)

- **Using the Figure,** TE. Students answer questions about the chapter opener photograph. (**GENERAL**)

- **Opening Activity,** Living Things Change, TE. Students brainstorm types of animals or plants that have changed over time. (**BASIC**)

Block 2

FOCUS (*5 minutes*)

- **Bellringer Transparency.** Use this transparency as students enter the classroom and find their seats. (**GENERAL**)

Copyright © by Holt, Rinehart and Winston. All rights reserved.

TEACHER RESOURCE PAGE

Lesson Plan continued

MOTIVATE (10 minutes)

- **Demonstration**, TE. Show students several different varieties of commercially important plants or animals, and ask students how these varieties originated. (**BASIC**)

TEACH (30 minutes s)

- **Teaching Transparency, Section Outline.** Use this transparency to give students a framework for the information in this section. (**GENERAL**)
- **Teaching Transparency, Darwin's Finches.** Use this transparency to link the changes in the finch's bills to the concept of descent with modification. (**GENERAL**)
- **Integrating Physics and Chemistry**, TE. Students design a hypothetical experiment to test Lamarck's theory of evolution.
- **Teaching Transparency, Two Rates of Progression.** Use this transparency to compare geometric and arithmetic progressions. Link geometric growth to Darwin's thinking about evolution. (**GENERAL**)

HOMEWORK

- **Directed Reading Worksheet, The Theory of Evolution by Natural Selection, CRF.** Students complete the exercises in this worksheet to help them understand the material as they read the section. (**BASIC**)
- **Active Reading Worksheet, The Theory of Evolution by Natural Selection, CRF.** Students read a passage related to the section topic and answer questions. (**GENERAL**)

Block 3

TEACH (35 minutes)

- **Skill Builder**, Graphing, TE. Students draw a graph with numerical data to reinforce the difference between geometric progression and arithmetic progression. (**GENERAL**)
- **Inclusion Strategies**, TE. Students describe and draw changes that would be necessary for animals represented by different animal crackers to evolve to live in a water environment and at the South Pole. (**GENERAL**)
- **Quick Lab**, Modeling Natural Selection, SE. Students play a timed game to model natural selection. (**GENERAL**)
- **Datasheets for In-Text Labs**, Modeling Natural Selection, CRF.
- **Exploring Further**, Punctuated Equilibrium, SE. Have students describe punctuated equilibrium and gradualism. Have students think about the similarities between these two models. (**GENERAL**)

TEACHER RESOURCE PAGE

Lesson Plan *continued*

CLOSE *(10 minutes)*

- **Reteaching,** TE. Point out that while in Indonesia, Wallace wrote a paper describing his idea about how evolution occurred and sent it to Darwin. Ask students to write a letter that Darwin could have written to Wallace in response. (**BASIC**)

- **Quiz,** TE. Students answer questions that review the section material. (**GENERAL**)

HOMEWORK

- **Alternative Assessment,** TE. Students make visual representations of the four major points that support Darwin's theory. (**GENERAL**)

- **Section Review,** SE. Assign questions 1–5 for review, homework, or quiz. (**GENERAL**)

- **Quiz, CRF.** This quiz consists of ten multiple choice and matching questions that review the section's main concepts. (**BASIC**) **Also in Spanish.**

Other Resource Options

- **Supplemental Reading, The Origin of Species, One-Stop Planner.** Students read the book and answer questions. (**ADVANCED**)

- **Group Activity,** Influences on Darwin, TE. Students research what scientific data and theories might have influence Darwin as he began conducting his investigations into the mechanism of evolution. (**ADVANCED**)

- **Internet Connect.** Students can research Internet sources about Natural Selection with SciLinks Code HX4128.

- **Internet Connect.** Students can research Internet sources about Theory of Evolution with SciLinks Code HX4175.

- **Internet Connect.** Students can research Internet sources about Fossil Record with SciLinks Code HX4088.

- **Internet Connect.** Students can research Internet sources about Evolution with SciLinks Code HX4074.

- **go.hrw.com.** For worksheets, videos, and other teaching aids related to this chapter, visit the HRW Web site and type in the keyword HX4 EVO.

- **CNN Student News.** Find the latest news, lesson plans, and activities related to important scientific events at **cnnstudentnews.com**.

- **CNN Science in the News, Video Segment 10 Galápagos.** This video segment is accompanied by a **Critical Thinking Worksheet**.

TEACHER RESOURCE PAGE

Lesson Plan

Section: Evidence of Evolution

Pacing

Regular Schedule: with lab(s): 3 days   without lab(s): 2 days
Block Schedule: with lab(s): 1 1/2 days   without lab(s): 1 day

Objectives

1. Describe how the fossil record supports evolution.
2. Summarize how biological molecules such as proteins and DNA are considered evidence of evolution.
3. Infer how comparing the anatomy and development of living species provides evidence of evolution.

National Science Education Standards Covered

UNIFYING CONCEPTS AND PROCESSES

UCP1: Systems, order, and organization

UCP2: Evidence, models, and explanation

UCP3: Change, constancy, and measurement

UCP4: Evolution and equilibrium

UCP5: Form and function

SCIENCE AS INQUIRY

SI1: Abilities necessary to do scientific inquiry

SI2: Understandings about scientific inquiry

HISTORY AND NATURE OF SCIENCE

HNS1: Science as a human endeavor

HNS2: Nature of scientific knowledge

HNS3: Historical perspectives

LIFE SCIENCE: THE CELL

LSCell1: Cells have particular structures that underlie their functions.

LSCell2: Most cell functions involve chemical reactions.

LSCell3: Cells store and use information to guide their functions.

LSCell4: Cell functions are regulated.

Copyright © by Holt, Rinehart and Winston. All rights reserved.

Holt Biology   The Theory of Evolution

TEACHER RESOURCE PAGE

Lesson Plan *continued*

LIFE SCIENCE: MOLECULAR BASIS OF HEREDITY

LSGene1: In all organisms, the instructions for specifying the characteristics of the organisms are carried in DNA.

LSGene3: Changes in DNA (mutations) occur spontaneously at low rates.

LIFE SCIENCE: BIOLOGICAL EVOLUTION

LSEvol1: Species evolve over time.

LSEvol2: The great diversity of organisms is the result of more than 3.5 billion years of evolution.

LSEvol3: Natural selection and its evolutionary consequences provide a scientific explanation for the fossil record of ancient life forms, as well as for the striking molecular similarities observed among the diverse species of living organisms.

LSEvol4: The millions of different species of plants, animals, and microorganisms that live on earth today are related by descent from common ancestors.

LSEvol5: Biological classifications are based on how organisms are related.

KEY
SE = Student Edition   **TE** = Teacher Edition
CRF = Chapter Resource File

Block 4

FOCUS *(5 minutes)*

- **Bellringer Transparency.** Use this transparency as students enter the classroom and find their seats. **(GENERAL)**

MOTIVATE *(10 minutes)*

- **Discussion/Question**, TE. Students listen to a passage from Darwin's *On the Origin of Species*, and describe what Darwin meant by this passage. **(GENERAL)**

TEACH *(30 minutes)*

- **Teaching Transparency, Section Outline.** Use this transparency to give students a framework for the information in this section. **(GENERAL)**

- **Teaching Transparency, Evidence of Whale Evolution.** Use this transparency to dicuss how whales are thought to have evolved from an ancestral line of four-legged mammals. Have volunteers describe similarities and differences in the bones of the animals shown. **(GENERAL)**

TEACHER RESOURCE PAGE
Lesson Plan continued

- **Using the Figure**, Figure 8, TE. Ask students how the backbones of the animals in the figure changed relative to the amount of time they spent in the water. (**GENERAL**)
- **Demonstration**, TE. Students study samples or pictures of different types of fossils and compare them to modern organisms. (**GENERAL**)

HOMEWORK

- **Directed Reading Worksheet, Evidence of Evolution, CRF.** Students complete the exercises in this worksheet to help them understand the material as they read the section. (**BASIC**)
- **Active Reading Worksheet, Evidence of Evolution, CRF.** Students read a passage related to the section topic and answer questions. (**GENERAL**)

Block 5

TEACH (25 minutes)

- **Teaching Transparency, Forelimbs of Vertebrates.** Use this transparency when discussing homologous structures. Have students describe the similarities in the structures of these forelimbs. (**GENERAL**)
- **Teaching Tip**, Making Mutations, TE. Students create mutations by substituting nitrogen bases in nucleotide sequences. (**GENERAL**)
- **Teaching Transparency, Hemoglobin Comparison.** Use this transparency to compare one human hemoglobin protein with the same protein in other species. Point out that the more similar organisms' hemoglobin proteins are the more recent the organisms' common ancestor is likely to have been. (**GENERAL**)

CLOSE (20 minutes)

- **Integrating Physics and Chemistry,** TE. Students analyze, review, and critique the evidence marshaled in support of the theory of evolution. (**GENERAL**)
- **Alternative Assessment**, TE. Tell students that the earliest phylogenetic trees were made using only evidence of morphological characteristics. Ask students why such trees might not reflect true evolutionary relationships. (**ADVANCED**)
- **Quiz**, TE. Students answer questions that review the section material. (**GENERAL**)

HOMEWORK

- **Reteaching**, TE. Students design a graphic organizer that describes two kinds of physical traits that can be used to support the theory of evolution. (**BASIC**)
- **Quick Lab, Comparing Limb Structure and Function, CRF.** Students compare the limbs of eight different animals, identify relationships between the structures, and look for evidence that these animals are related to one another (**BASIC**)
- **Section Review**, SE. Assign questions 1–4 for review, homework, or quiz. (**GENERAL**)

TEACHER RESOURCE PAGE

Lesson Plan continued

- **Quiz, CRF.** This quiz consists of ten multiple choice and matching questions that review the section's main concepts. (**BASIC**) **Also in Spanish.**

Optional Block

LAB *(45 minutes)*

- **Skills Practice Lab, Melanism in Insects, CRF.** Students simulate how successfully predators locate prey in different environments. Then they relate changes in a population of beetles with two color variations to changes in the environment. (**GENERAL**)

Other Resource Options

- **Supplemental Reading, The Origin of Species, One-Stop Planner.** Students read the book and answer questions. (**ADVANCED**)
- **Internet Connect.** Students can research Internet sources about Natural Selection with SciLinks Code HX4128.
- **Internet Connect.** Students can research Internet sources about Theory of Evolution with SciLinks Code HX4175.
- **Internet Connect.** Students can research Internet sources about Fossil Record with SciLinks Code HX4088.
- **Internet Connect.** Students can research Internet sources about Evolution with SciLinks Code HX4074.
- **go.hrw.com.** For worksheets, videos, and other teaching aids related to this chapter, visit the HRW Web site and type in the keyword HX4 EVO.
- **CNN Student News.** Find the latest news, lesson plans, and activities related to important scientific events at **cnnstudentnews.com**.

TEACHER RESOURCE PAGE
Lesson Plan

Section: Examples of Evolution

Pacing
Regular Schedule: with lab(s): 3 days   without lab(s): 2 days
Block Schedule: with lab(s): 1 1/2 days   without lab(s): 1 day

Objectives
1. Identify four elements on the process of natural selection.
2. Describe how natural selection has affected the bacteria that cause tuberculosis.
3. Relate natural selection to the beak size of finches.
4. Summarize the process of species formation.

National Science Education Standards Covered
UNIFYING CONCEPTS AND PROCESSES

UCP1: Systems, order, and organization

UCP2: Evidence, models, and explanation

UCP4: Evolution and equilibrium

UCP5: Form and function

SCIENCE AS INQUIRY

SI1: Abilities necessary to do scientific inquiry

SI2: Understandings about scientific inquiry

HISTORY AND NATURE OF SCIENCE

HNS1: Science as a human endeavor

HNS2: Nature of scientific knowledge

HNS3: Historical perspectives

LIFE SCIENCE: MOLECULAR BASIS OF HEREDITY

LSGene3: Changes in DNA (mutations) occur spontaneously at low rates.

LIFE SCIENCE: BIOLOGICAL EVOLUTION

LSEvol1: Species evolve over time.

LSEvol2: The great diversity of organisms is the result of more than 3.5 billion years of evolution.

TEACHER RESOURCE PAGE

Lesson Plan continued

LSEvol3: Natural selection and its evolutionary consequences provide a scientific explanation for the fossil record of ancient life forms, as well as for the striking molecular similarities observed among the diverse species of living organisms.

LSEvol4: The millions of different species of plants, animals, and microorganisms that live on earth today are related by descent from common ancestors.

LSEvol5: Biological classifications are based on how organisms are related.

KEY
SE = Student Edition   TE = Teacher Edition
CRF = Chapter Resource File

Block 6

FOCUS (5 minutes)

- **Bellringer Transparency.** Use this transparency as students enter the classroom and find their seats. (**GENERAL**)

MOTIVATE (10 minutes)

- **Discussion/Question**, TE. Discuss with students the great variation of life forms within the phylum Arthropoda. (**GENERAL**)

TEACH (30 minutes)

- **Teaching Transparency, Section Outline.** Use this transparency to give students a framework for the information in this section. (**GENERAL**)
- **Inclusion Strategies**, TE. Students make a chart of the difference between how polar bears and brown bears live and survive in their environments. (**BASIC**)
- **Teaching Transparency, Beak Size Variation.** Use this transparency to discuss how ozone formed as cyanobacteria added oxygen to the atmosphere. (**GENERAL**)
- **Teaching Transparency, Mating Activity in Raven Species.** Use this transparency to discuss how ozone formed as cyanobacteria added oxygen to the atmosphere. (**GENERAL**)
- **Teaching Tip**, Selection Pressure and Rates of Evolution, TE. Lead a discussion on natural selection. Tell students that strong selection pressure occurs if an organism with a particular trait dies before reaching reproductive age. (**GENERAL**)

HOMEWORK

- **Directed Reading Worksheet, Examples of Evolution, CRF.** Students complete the exercises in this worksheet to help them understand the material as they read the section. (**BASIC**)
- **Active Reading Worksheet, Examples of Evolution, CRF.** Students read a passage related to the section topic and answer questions. (**GENERAL**)

Copyright © by Holt, Rinehart and Winston. All rights reserved.

Holt Biology   The Theory of Evolution

TEACHER RESOURCE PAGE

Lesson Plan *continued*

- **Group Activity**, Examples of Evolution, TE. Students research examples of evolution that show evolutionary change within a human lifespan. (**ADVANCED**)

Block 7

TEACH *(35 minutes)*

- **Group Activity**, Modeling Natural Selection, TE. Students learn about selective pressure by completing an exercise with paper and aluminum foil. (**BASIC**)
- **Teaching Tip**, Sequential Diagram of Speciation, TE. Students make a graphic organizer that shows each step of speciation in the proper order. (**GENERAL**)
- **Math Lab**, Analyzing Change in Lizard Populations, SE. Students analyze a data table and use the information to predict changes that might occur. (**GENERAL**)
- **Datasheets for In-Text Labs**, Analyzing Change in Lizard Populations, CRF.

CLOSE *(10 minutes)*

- **Reteaching**, TE. Students suggest why sharks and alligators have not changed much over millions of years. (**BASIC**)
- **Quiz**, TE. Students answer questions that review the section material. (**GENERAL**)

HOMEWORK

- **Alternative Assessment**, TE. Students make a labeled diagram showing the process of natural selection using species of their choice. (**GENERAL**)
- **Science Skills Worksheet**, CRF. Students analyze a story and identify how the information applies to the process of natural selection. (**GENERAL**)
- **Section Review**, SE. Assign questions 1–6 for review, homework, or quiz. (**GENERAL**)
- **Quiz**, CRF. This quiz consists of ten multiple choice and matching questions that review the section's main concepts. (**BASIC**) Also in Spanish.
- **Modified Worksheet, One-Stop Planner.** This worksheet has been specially modified to reach struggling students. (**BASIC**)
- **Critical Thinking Worksheet**, CRF. Students answer analogy-based questions that review the section's main concepts and vocabulary. (**ADVANCED**)

Optional Block

LAB *(45 minutes)*

- **Exploration Lab**, Modeling Natural Selection, SE. Students model the selection of favorable traits in a new generation using a paper model of a bird as their study species. (**GENERAL**)
- **Datasheets for In-Text Labs**, Modeling Natural Selection, CRF.

Copyright © by Holt, Rinehart and Winston. All rights reserved.

Holt Biology · The Theory of Evolution

TEACHER RESOURCE PAGE

Lesson Plan *continued*

Other Resource Options

- **Integrating Physics and Chemistry**, TE. Students compare and contrast the field observations and conclusions of Lack and the Grants.
- **Internet Connect.** Students can research Internet sources about Natural Selection with SciLinks Code HX4128.
- **Internet Connect.** Students can research Internet sources about Theory of Evolution with SciLinks Code HX4175.
- **Internet Connect.** Students can research Internet sources about Fossil Record with SciLinks Code HX4088.
- **Internet Connect.** Students can research Internet sources about Evolution with SciLinks Code HX4074.
- **go.hrw.com.** For worksheets, videos, and other teaching aids related to this chapter, visit the HRW Web site and type in the keyword HX4 EVO.
- **CNN Student News.** Find the latest news, lesson plans, and activities related to important scientific events at **cnnstudentnews.com**.
- **CNN Science in the News, Video Segment 10 Galápagos.** This video segment is accompanied by a **Critical Thinking Worksheet**.

TEACHER RESOURCE PAGE
Lesson Plan

End-of-Chapter Review and Assessment

Pacing
Regular Schedule: 2 days
Block Schedule: 1 day

KEY
SE = Student Edition   TE = Teacher Edition
CRF = Chapter Resource File

Block 8
REVIEW *(45 minutes)*

- **Study Zone,** SE. Use the Study Zone to review the Key Concepts and Key Terms of the chapter and prepare students for the Performance Zone questions. (**GENERAL**)

- **Performance Zone,** SE. Assign questions to review the material for this chapter. Use the assignment guide to customize review for sections covered. (**GENERAL**)

- **Teaching Transparency, Concept Mapping.** Use this transparency to review the concept map for this chapter. (**GENERAL**)

Block 9
ASSESSMENT *(45 minutes)*

- **Chapter Test, The Theory of Evolution, CRF.** This test contains 20 multiple choice and matching questions keyed to the chapter's objectives. (**GENERAL**) **Also in Spanish.**

- **Chapter Test, The Theory of Evolution, CRF.** This test contains 25 questions of various formats, each keyed to the chapter's objectives. (**ADVANCED**)

- **Modified Chapter Test, One-Stop Planner.** This test has been specially modified to reach struggling students. (**BASIC**)

Other Resource Options

- **Vocabulary Review Worksheet, CRF.** Use this worksheet to review the chapter vocabulary. (**GENERAL**) **Also in Spanish.**

- **Test Prep Pretest, CRF.** Use this pretest to review the main content of the chapter. Each question is keyed to a section objective. (**GENERAL**) **Also in Spanish.**

- **Test Item Listing for ExamView® Test Generator, CRF.** Use the Test Item Listing to identify questions to use in a customized homework, quiz, or test.

- **ExamView® Test Generator, One-Stop Planner.** Create a customized homework, quiz, or test using the HRW Test Generator program.

Copyright © by Holt, Rinehart and Winston. All rights reserved.

TEST ITEM LISTING
The Theory of Evolution

TRUE/FALSE

1. ____ Species that have evolved from a common ancestor should have certain characteristics in common.
   Answer: True   Difficulty: I   Section: 1   Objective: 1

2. ____ In his *Essay on the Principle of Population,* Malthus said humans were the only population that could continue to grow in size indefinitely.
   Answer: False   Difficulty: I   Section: 1   Objective: 1

3. ____ Darwin observed that the plants and animals of the Galápagos Islands were the same as those on islands off the coast of Africa with similar environments.
   Answer: False   Difficulty: I   Section: 1   Objective: 1

4. ____ The book *Principles of Geology* by Charles Lyell described how changes in land formations can cause species to evolve.
   Answer: False   Difficulty: I   Section: 1   Objective: 1

5. ____ The inheritance of acquired characteristics was one mechanism of evolution supported by Darwin.
   Answer: False   Difficulty: I   Section: 1   Objective: 2

6. ____ Natural selection can cause the spread of an advantageous adaptation throughout a population over time.
   Answer: True   Difficulty: I   Section: 1   Objective: 2

7. ____ The two major ideas that Darwin presented in *The Origin of the Species* were that evolution occurred and that natural selection was its mechanism.
   Answer: True   Difficulty: I   Section: 1   Objective: 3

8. ____ The theory of evolution states that species change over time.
   Answer: True   Difficulty: I   Section: 1   Objective: 3

9. ____ Natural selection causes allele frequencies within populations to remain the same.
   Answer: False   Difficulty: I   Section: 1   Objective: 3

10. ____ Punctuated gradualism refers to the hypothesis that evolution occurs only in short periods of time.
    Answer: False   Difficulty: I   Section: 1   Objective: 4

11. ____ Two hypotheses suggested about the rate at which evolution proceeds are gradualism and punctuated equilibrium.
    Answer: True   Difficulty: I   Section: 1   Objective: 4

12. ____ The fossil record suggests that species have become less complex over time.
    Answer: False   Difficulty: I   Section: 2   Objective: 1

13. ____ The theory of evolution predicts that genes will accumulate more alterations in their nucleotide sequences over time.
    Answer: True   Difficulty: I   Section: 2   Objective: 2

14. ____ Evidence for evolution occurs only in the fossil record.
    Answer: False   Difficulty: I   Section: 2   Objective: 2

15. ____ The human forelimb and the bat forelimb are homologous structures.
    Answer: True   Difficulty: I   Section: 2   Objective: 3

TEST ITEM LISTING, *continued*

16. ____ Early in development, human embryos and the embryos of all other vertebrates are strikingly similar.
    Answer: True   Difficulty: I   Section: 2   Objective: 3

17. ____ The way an embryo develops is not important in determining the evolutionary history of a species.
    Answer: False   Difficulty: I   Section: 2   Objective: 3

18. ____ The environment dictates only the direction and extent of evolution.
    Answer: True   Difficulty: I   Section: 3   Objective: 1

19. ____ The bacteria that cause tuberculosis have been unaffected by natural selection.
    Answer: False   Difficulty: I   Section: 3   Objective: 2

20. ____ Mutant bacteria that cause tuberculosis were selected against by natural selection mechanisms.
    Answer: False   Difficulty: I   Section: 3   Objective: 2

21. ____ A mutation in the bacteria that cause tuberculosis made them resistant to antibiotics.
    Answer: True   Difficulty: I   Section: 3   Objective: 2

22. ____ The environment selects which organisms will survive and reproduce by presenting challenges that only individuals with particular traits can meet.
    Answer: True   Difficulty: I   Section: 3   Objective: 2

23. ____ When food is plentiful, there is little selective pressure on the beaks of finches.
    Answer: True   Difficulty: I   Section: 3   Objective: 3

24. ____ When food is scarce, there is little selective pressure on the beaks of finches.
    Answer: False   Difficulty: I   Section: 3   Objective: 3

25. ____ When food is scarce, the number of different beak shapes of finches increases.
    Answer: True   Difficulty: I   Section: 3   Objective: 3

26. ____ The accumulation of differences between species or populations is called convergence.
    Answer: False   Difficulty: I   Section: 3   Objective: 4

27. ____ Within populations, divergence leads to speciation.
    Answer: True   Difficulty: I   Section: 3   Objective: 4

MULTIPLE CHOICE

28. Darwin thought that the plants and animals of the Galápagos Islands were similar to those of the nearby coast of South America because
    a. their ancestors had migrated from South America to the Galápagos Islands.
    b. they had all been created by God to match their habitat.
    c. the island organisms had the same nucleotide sequences in their DNA as the mainland organisms.
    d. he found fossils, proving that the animals and plants had common ancestors.
    Answer: A   Difficulty: I   Section: 1   Objective: 1

29. Darwin conducted much of his research on
    a. the Samoan Islands.   c. the Hawaiian Islands.
    b. Manhattan Island.   d. the Galápagos Islands.
    Answer: D   Difficulty: I   Section: 1   Objective: 1

30. Which of the following describes a population?
    a. dogs and cats living in Austin, Texas
    b. four species of fish living in a pond
    c. dogwood trees in Middletown, Connecticut
    d. roses and tulips in a garden
    Answer: C   Difficulty: I   Section: 1   Objective: 1

31. Natural selection is the process by which
    a. the age of selected fossils is calculated.
    b. organisms with traits well suited to their environment survive and reproduce at a greater rate than less well-adapted organisms in the same environment.
    c. acquired traits are passed on from one generation to the next.
    d. All of the above
    Answer: B   Difficulty: I   Section: 1   Objective: 2

32. Natural selection could *not* occur without
    a. genetic variation in species.
    b. environmental changes.
    c. competition for unlimited resources.
    d. gradual warming of Earth.
    Answer: A   Difficulty: I   Section: 1   Objective: 2

33. Natural selection causes
    a. changes in the environment.
    b. plants and animals to produce more offspring than can survive.
    c. changes in the frequency of certain alleles in a population.
    d. All of the above
    Answer: C   Difficulty: I   Section: 1   Objective: 2

34. The process by which a species becomes better suited to its environment is known as
    a. accommodation.
    b. variation.
    c. adaptation.
    d. selection.
    Answer: C   Difficulty: I   Section: 1   Objective: 3

35. According to Darwin, evolution occurs
    a. by chance.
    b. during half-life periods of 5,730 years.
    c. because of natural selection.
    d. rapidly.
    Answer: C   Difficulty: I   Section: 1   Objective: 3

36. Organisms well suited to their environment
    a. reproduce at a greater rate than those less suited to the same environment.
    b. are always larger than organisms less suited to that environment.
    c. always live longer than organisms less suited to that environment.
    d. need less food than organisms less suited to that environment.
    Answer: A   Difficulty: I   Section: 1   Objective: 3

37. When Darwin published his theory of evolution, he included all of the following ideas *except*
    a. the idea that species change slowly over time.
    b. the idea that some organisms become less suited to their environment than others.
    c. Mendel's ideas about genetics.
    d. the idea that some organisms reproduce at a greater rate than others.
    Answer: C   Difficulty: I   Section: 1   Objective: 3

38. The major idea that Darwin presented in his book *The Origin of Species* was that
    a. species changed over time and never competed with each other.
    b. animals changed, but plants remained the same.
    c. elephants and bacteria changed constantly.
    d. species changed over time by natural selection.
    Answer: D   Difficulty: I   Section: 1   Objective: 3

TEST ITEM LISTING, continued

39. The hypothesis that evolution occurs at a slow, constant rate is known as
    a. gradualism.
    b. slow motion.
    c. natural selection.
    d. adaptation.

    Answer: A Difficulty: I Section: 1 Objective: 4

40. The hypothesis that evolution occurs at an irregular rate through geologic time is known as
    a. directional evolution.
    b. directional equilibrium.
    c. punctuated equilibrium.
    d. punctuated evolution.

    Answer: C Difficulty: I Section: 1 Objective: 4

The diagrams below represent bones in the limbs of fossil horses and modern horses.

60 million years ago   modern

41. Refer to the illustration above. The fossils indicate that horse evolution probably has taken place
    a. rapidly.
    b. in only one place on Earth.
    c. gradually.
    d. five times by the process of punctuated equilibrium.

    Answer: C Difficulty: II Section: 2 Objective: 1

42. Which of the following are examples of fossils?
    a. shells or old bones
    b. any traces of dead organisms
    c. footprints of human ancestors, insects trapped in tree sap, and animals buried in tar
    d. All of the above

    Answer: D Difficulty: I Section: 2 Objective: 1

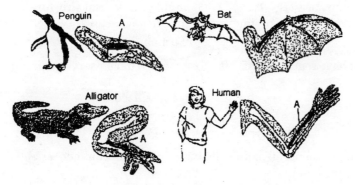

43. Refer to the illustration above. An analysis of DNA from these organisms would indicate that
    a. they have identical DNA.
    b. they all have pharyngeal pouches.
    c. their nucleotide sequences show many similarities.
    d. they all have the same number of chromosomes.

    Answer: C Difficulty: II Section: 2 Objective: 2

TEST ITEM LISTING, continued

44. Refer to the illustration above. The similarity of these structures suggests that the organisms
    a. have a common ancestor.
    b. all grow at different rates.
    c. evolved slowly.
    d. live for a long time.

    Answer: A   Difficulty: II   Section: 2   Objective: 3

45. Refer to the illustration above. The bones labeled A are known as
    a. vestigial structures.
    b. sequential structures.
    c. homologous structures.
    d. fossil structures.

    Answer: C   Difficulty: II   Section: 2   Objective: 3

46. The theory of evolution predicts that
    a. closely related species will show similarities in nucleotide sequences.
    b. if species have changed over time, their genes should have changed.
    c. closely related species will show similarities in amino acid sequences.
    d. All of the above

    Answer: D   Difficulty: I   Section: 2   Objective: 2

47. The occurrence of the same blood protein in a group of species provides evidence that these species
    a. evolved in the same habitat.
    b. evolved in different habitats.
    c. descended from a common ancestor.
    d. descended from different ancestors.

    Answer: C   Difficulty: I   Section: 2   Objective: 2

48. Evidence for evolution includes all of the following *except*
    a. punctuated sedimentation.
    b. similarities and differences in protein and DNA sequences between organisms.
    c. the fossil record.
    d. homologous structures.

    Answer: A   Difficulty: I   Section: 2   Objective: 3

49. Which of the following is a vestigial structure?
    a. the human tailbone
    b. the bill of a finch
    c. flower color
    d. fossil cast

    Answer: A   Difficulty: I   Section: 2   Objective: 3

50. Homologous structures in organisms suggest that the organisms
    a. have a common ancestor.
    b. must have lived at different times.
    c. have a skeletal structure.
    d. are now extinct.

    Answer: A   Difficulty: I   Section: 2   Objective: 3

51. Structures that have reduced in size because they no longer serve an important function are called
    a. inorganic.
    b. mutated.
    c. fossilized.
    d. vestigial.

    Answer: D   Difficulty: I   Section: 2   Objective: 3

52. A human embryo exhibits all of the following during development *except*
    a. pharyngeal pouches.
    b. a bony tail.
    c. fins.
    d. a coat of fine fur.

    Answer: C   Difficulty: I   Section: 2   Objective: 3

TEST ITEM LISTING, continued

53. vestigial structures : macroevolution ::
    a. homologous structures : common ancestry
    b. common ancestry : fossils
    c. common ancestry : rock
    d. homologous structures : unrelated species
    Answer: A   Difficulty: II   Section: 2   Objective: 3

54. Populations of the same species living in different places
    a. do not vary.
    b. always show balancing selection.
    c. have a half-life in relation to the size of the population.
    d. become increasingly different as each becomes adapted to its own environment.
    Answer: D   Difficulty: I   Section: 3   Objective: 1

55. Scarcity of resources and a growing population are most likely to result in
    a. homology.   c. competition.
    b. protective coloration.   d. convergent evolution.
    Answer: C   Difficulty: I   Section: 3   Objective: 1

56. Since natural resources are limited, all organisms
    a. must migrate to new habitats.   c. display vestigial structures.
    b. face a constant struggle for existence. d. have a species half-life.
    Answer: B   Difficulty: I   Section: 3   Objective: 1

57. A change in the frequency of a particular gene in one direction in a population is a result of
    a. natural selection.   c. chromosome drift.
    b. acquired variation.   d. balancing selection.
    Answer: A   Difficulty: I   Section: 3   Objective: 1

58. struggle for survival : competition ::
    a. time : environment   c. trait : time
    b. survival of the fittest : best traits d. environment : traits
    Answer: B   Difficulty: II   Section: 3   Objective: 1

59. *Mycobacterium tuberculosis*
    a. always responds to antibiotics.
    b. can mutate and become resistant to antibiotics.
    c. is a harmless organism that normally occurs in human lungs.
    d. has never responded to antibiotics.
    Answer: B   Difficulty: I   Section: 3   Objective: 2

60. The lung disease tuberculosis
    a. kills more adults than any other infectious disease.
    b. is easily treated with rifampin and isoniazid.
    c. is caused by an unknown organism.
    d. usually affects only children.
    Answer: A   Difficulty: I   Section: 3   Objective: 2

61. The mutation that made *Mycobacterium tuberculosis* resistant to antibiotics involved
    a. a missing chromosome.
    b. an extra gene.
    c. a single base change.
    d. None of the above
    Answer: C   Difficulty: I   Section: 3   Objective: 2

TEST ITEM LISTING, continued

62. Rifampin, the antobiotic commonly used to treat tuberculosis, acts by
    a. mutating bacterial RNA.
    b. preventing bacteria from dividing.
    c. mutating bacterial polymerase genes.
    d. preventing bacterial mRNA transcription.
    Answer: D   Difficulty: I   Section: 3   Objective: 2

63. The finches that Darwin studied differed in the shape of their beaks. According to Darwin, the finches probably
    a. all had a common ancestor.
    b. had been created by design that way.
    c. were descended from similar birds in Africa.
    d. ate the same diet.
    Answer: A   Difficulty: I   Section: 3   Objective: 3

64. Beak shape in finches is affected by
    a. the number of predators in the area.   c. the color of the finch.
    b. the size of the finch.   d. the availability of food.
    Answer: D   Difficulty: I   Section: 3   Objective: 3

65. In order to fit into their habitat, the Galapágos finches had
    a. not changed.   c. evolved.
    b. been created as superior birds.   d. None of the above
    Answer: C   Difficulty: I   Section: 3   Objective: 3

66. The accumulation of differences between species or populations is called
    a. gradualism.   c. divergence.
    b. adaptation.   d. differentiation.
    Answer: C   Difficulty: I   Section: 3   Objective: 4

67. Which of the following statements is *not* true about members of subspecies?
    a. Members of different subspecies are not yet different enough to belong to separate species.
    b. Members of one subspecies cannot interbreed with members of any other such group.
    c. Subspecies often become increasingly different in response to their environment.
    d. Divergence between subspecies occurs because natural selection favors different survival strategies in different environments.
    Answer: B   Difficulty: I   Section: 3   Objective: 4

68. New species form
    a. when subspecies diverge more and more.
    b. because of natural selection.
    c. when members of the same species become adapted to new environments.
    d. All of the above
    Answer: D   Difficulty: I   Section: 3   Objective: 4

69. Populations of the same species that differ genetically because they have adapted to different living conditions are known as
    a. selected populations.   c. genetic populations.
    b. subspecies.   d. genetic races.
    Answer: B   Difficulty: I   Section: 3   Objective: 4

COMPLETION

70. A change in species over time is called _____.
    Answer: evolution   Difficulty: I   Section: 1   Objective: 1

TEST ITEM LISTING, continued

71. Charles Darwin sailed for five years on a ship named _____ _____.

   Answer: H.M.S. *Beagle*   Difficulty: I   Section: 1   Objective: 1

72. The process by which organisms with traits well suited to an environment survive and reproduce at a greater rate than organisms less suited for that environment is called _____ _____.

   Answer: natural selection   Difficulty: I   Section: 1   Objective: 2

73. Natural selection leads to changes in both the physical appearance and the _____ _____ of a species.

   Answer: genetic makeup   Difficulty: I   Section: 1   Objective: 2

74. Published in 1859, Charles Darwin's book, _____ _____ _____ _____, changed biology forever.

   Answer: *The Origin of Species*   Difficulty: I   Section: 1   Objective: 3

75. A species that has disappeared permanently is said to be _____.

   Answer: extinct   Difficulty: I   Section: 1   Objective: 3

76. The most direct evidence that evolution has occurred comes from _____.

   Answer: fossils   Difficulty: I   Section: 2   Objective: 1

77. Closely related species show more _____ in nucleotide sequences than distantly related species.

   Answer: similarities   Difficulty: I   Section: 2   Objective: 2

78. Homologous structures are similar because they are inherited from a common _____.

   Answer: ancestor   Difficulty: I   Section: 2   Objective: 3

79. Eyes in a blind salamander are an example of a type of organ known as _____.

   Answer: vestigial   Difficulty: I   Section: 2   Objective: 3

80. Because they are inherited from a common ancestor, _____ structures are similar.

   Answer: homologous   Difficulty: I   Section: 2   Objective: 3

81. Evolution that occurs at a constant rate is the hypothesis called _____.

   Answer: gradualism   Difficulty: I   Section: 2   Objective: 4

82. The raw material for natural selection is _____ _____.

   Answer: genetic variation   Difficulty: I   Section: 3   Objective: 1

83. According to Darwin, the _____ determines the rate at which organisms survive and reproduce.

   Answer: environment   Difficulty: I   Section: 3   Objective: 2

84. Some bacteria have developed _____ _____ through the process of natural selection.

   Answer: antibiotic resistance   Difficulty: I   Section: 3   Objective: 2

85. Mutant *Mycobacterium tuberculosis* is more dangerous than the normal strain because it is resistant to _____.

   Answer: antibiotics   Difficulty: I   Section: 3   Objective: 2

86. The mutant form of disease-causing bacteia becoming predominant is a result of _____ _____.

   Answer: natural selection   Difficulty: I   Section: 3   Objective: 2

TEST ITEM LISTING, continued

87. Darwin's observations of finches led him to believe that there was a close correlation between beak shape and _____ source.
    Answer: food   Difficulty: I   Section: 3   Objective: 3

88. The availability of food supply affects the number of different _____ shapes in finches.
    Answer: beak   Difficulty: I   Section: 3   Objective: 3

ESSAY

89. Why did Darwin believe that the finches he observed and collected in the Galápagos Islands shared a common ancestor?
    Answer:
    Although there were differences among these finch species, all the species also had many traits in common. The main similarities among these species led Darwin to conclude that they had a common ancestor.
    Difficulty: II   Section: 1   Objective: 1

90. In comparing two species that look very different, how could a comparison of the species' genes contribute to an understanding of their evolutionary relationship?
    Answer:
    Studying the species' genes would provide much more information than could be obtained by simply observing the physical appearance of the species. If the species had many genes in common, they would likely be more closely related than their physical appearance would suggest. If the species did not have many genes in common, this information would tend to strengthen the argument that the species were not closely related.
    Difficulty: III   Section: 2   Objective: 2

91. You are a biologist accompanying some other scientists on an expedition in a region that has not been studied intensively. In your explorations, you come across a colony of small vertebrates that do not look familiar to you. After conducting electronic searches of worldwide databases, you arrive at the tentative conclusion that this organism has never been observed before. Now your job is to determine what kind of vertebrate it is by identifying its closest relatives. Identify three types of data that you would collect and describe how you would use these data to draw your conclusions.
    Answer:
    a. Analysis of anatomical structures and comparison of these to similar structures of other vertebrates is one type of data that should be collected. For example, the bones composing the forelimb of the organism could be compared to the forelimbs of other vertebrates. Those vertebrates having the greatest number of similar (homologous) anatomical structures to those of your organism could be presumed to be its closest relatives.
    b. Analysis of the DNA and/or a protein and comparison of this material to that of other vertebrates could also be studied. For example, DNA hybridization studies could be conducted with the organism and other vertebrates. Or, an analysis of the cytochrome c of the organism in comparison to the cytochrome c of other vertebrates could be done. Those vertebrates having the fewest differences in sequences of DNA and/or proteins from the organism could be presumed to be its closest relatives.
    c. Analysis of embryonic development and comparison of structures present at different stages and the pattern of development with the structures and patterns of other vertebrates would be a third type of data collected. For example, an analysis could be made of the persistence of a particular trait until late in embryonic development. This analysis could be compared to the persistence of the same trait in the embryos of other vertebrates. Those vertebrates having the greatest similarity in structures present and pattern of development could be presumed to be its closest relatives.
    Difficulty: III   Section: 2   Objective: 3

TEST ITEM LISTING, continued

92. Why is competition among individuals of the same species generally so intense?
    Answer:
    Individuals of the same species require the same resources for survival. Since resources are generally limited, only those individuals able to secure sufficient amounts of such resources will survive.
    Difficulty: II   Section: 3   Objective: 1

93. An agricultural plot of land is sprayed with a very powerful insecticide to destroy harmful insects. Nevertheless, many of the same species of insects are present on the land the following year. How might the theory of evolution account for this phenomenon?
    Answer:
    A part of the theory of evolution states that genetic variation exists within a species. A small percentage of the insects exposed to the insecticide might have been immune or capable of detoxifying the substance. They survived and produced offspring that were also resistant to the insecticide.
    Difficulty: III   Section: 3   Objective: 1

94. What role does the environment play in natural selection?
    Answer:
    Those organisms that have traits best suited to the environment most successfully survive and reproduce.
    Difficulty: II   Section: 3   Objective: 2

95. Suppose that you are a zoologist studying birds on a group of islands. You have just discovered four species of birds that have never been seen before. Each species is on a separate island. The birds are identical to one another except for the shape of their beaks. How can you explain their similarities and differences?
    Answer:
    It is likely that the four species evolved from a common ancestor, with each species adapting to the conditions on its island. The differences in beak shape may be the result of differences in available food among the islands. Each bird species adapted to the food that was available on its island.
    Difficulty: III   Section: 3   Objective: 4